Association for Women in Mathematics Series

Volume 19

Series Editor
Kristin Lauter
Microsoft Research
Redmond, Washington, USA

Association for Women in Mathematics Series

Focusing on the groundbreaking work of women in mathematics past, present, and future, Springer's Association for Women in Mathematics Series presents the latest research and proceedings of conferences worldwide organized by the Association for Women in Mathematics (AWM). All works are peer-reviewed to meet the highest standards of scientific literature, while presenting topics at the cutting edge of pure and applied mathematics. Since its inception in 1971, The Association for Women in Mathematics has been a non-profit organization designed to help encourage women and girls to study and pursue active careers in mathematics and the mathematical sciences and to promote equal opportunity and equal treatment of women and girls in the mathematical sciences. Currently, the organization represents more than 3000 members and 200 institutions constituting a broad spectrum of the mathematical community, in the United States and around the world.

More information about this series at http://www.springer.com/series/13764

Jennifer S. Balakrishnan • Amanda Folsom
Matilde Lalín • Michelle Manes
Editors

Research Directions in Number Theory

Women in Numbers IV

ASSOCIATION FOR
WOMEN IN MATHEMATICS

 Springer

Editors

Jennifer S. Balakrishnan
Department of Mathematics and Statistics
Boston University
Boston, MA, USA

Amanda Folsom
Department of Mathematics and Statistics
Amherst College
Amherst, MA, USA

Matilde Lalín
Département de mathématiques et de
statistique
Université de Montréal
Montréal, QC, Canada

Michelle Manes
Department of Mathematics
University of Hawai'i
Honolulu, HI, USA

ISSN 2364-5733   ISSN 2364-5741   (electronic)
Association for Women in Mathematics Series
ISBN 978-3-030-19480-2   ISBN 978-3-030-19478-9   (eBook)
https://doi.org/10.1007/978-3-030-19478-9

Mathematics Subject Classification: 05C25, 14G50, 11G20

This Springer imprint is published by the registered company Springer Nature Switzerland AG.
The registered company address is: Gewerbestrasse 11, 6330 Cham, Switzerland

Fig. 1 Conference photo, courtesy of Banff International Research Station

Preface

This volume is a compilation of research and survey papers in number theory, written by members of the Women in Numbers (WIN) network, principally by the collaborative research groups formed at Women in Numbers 4 (WIN4), a conference at the Banff International Research Station in Banff, Alberta, on August 14–18, 2017.

The WIN conference series began in 2008. The series introduced a novel research-mentorship model: women at all career stages, from graduate students to senior members of the community, joined forces to work in focused research groups on cutting-edge projects designed and led by experienced researchers. This model has proven so successful that to date there are nearly 20 research networks for women in mathematics, each of which holds Research Collaboration Conferences for Women as well as other conferences, workshops, special sessions, and symposia. The Association for Women in Mathematics (AWM), funded by the National Science Foundation ADVANCE program, is now supporting and researching the effectiveness of this research-mentorship model (https://awmadvance.org/rccws/).

The goals for WIN4 were to generate research in significant topics in number theory; to broaden the research programs of women and gender minorities working in number theory, especially pre-tenure; to train graduate students and postdocs in number theory by providing experience with collaborative research and the publication process; to strengthen and extend a research network of potential collaborators in number theory and related fields; to enable faculty at small colleges to participate actively in research activities including mentoring graduate students and postdocs; and to highlight research activities of women in number theory.

The majority of the week was devoted to research activities. Before the conference, the participants were organized into nine project groups by research interest and asked to learn background for their project topics. During the workshop, the group leaders gave short talks to all the participants introducing their general areas of research and their groups' projects. On the final day, the group members described their progress and shared their plans to complete the work.

Forty-two mathematicians attended the WIN4 workshop, which was organized by Editors Balakrishnan and Manes along with Chantal David (Concordia University) and Bianca Viray (University of Washington).

The editors solicited contributions from the working groups at the WIN4 workshop and sought additional articles through the Women in Numbers Network (mailing list and web site). All submissions to this volume were sent to anonymous referees, who assessed the work as correct and worthwhile contributions to these proceedings. This volume is the sixth proceedings released after a WIN conference.

The articles collected here span algebraic, analytic, and computational areas of number theory, including topics such as elliptic and hyperelliptic curves, mock modular forms, arithmetic dynamics, and cryptographic applications. Several papers in this volume stem from collaborations between authors with different mathematical backgrounds, allowing the group to tackle a problem using multiple perspectives and tools. In what follows, we highlight some connections between the articles in this volume and also the subjects covered.

Bridging the areas of number theory and cryptography is the article *Ramanujan Graphs in Cryptography* (Costache et al.). From the perspective of both subjects, this paper studies the security of a proposal for post-quantum cryptography.

Four papers in this volume surround computational aspects of curves, varieties, and surfaces. *Computational Aspects of Supersingular Elliptic Curves* (Bank et al.) studies the problem of generating the endomorphism ring of a supersingular elliptic curve by two cycles in ℓ-isogeny graphs, while *Chabauty-Coleman Experiments on Genus Three Hyperelliptic Curves* (Balakrishnan et al.) describes a computation of rational points on genus three hyperelliptic curves defined over $\mathbb{Q}$ whose Jacobians have Mordell-Weil rank 1. *Weierstrass Equations for the Elliptic Fibrations of a K3 Surface* (Lecacheux) concludes a study of the classification of elliptic fibrations of a singular $K3$ surface by giving all Weierstrass equations. Lastly, within this theme, *Newton Polygons of Cyclic Covers of the Projective Line* (Li et al.) applies the Shimura-Taniyama method for computing the Newton polygon of an abelian variety with complex multiplication to cyclic covers of the projective line branched at three points and produces multiple new examples.

Arithmetic dynamics is another subject explored in multiple papers and from different standpoints: *Arithmetic Dynamics and Galois Representations* (Juul et al.) proves a version of Jones' conjectures on the arboreal representation of a degree two rational map, and *Dessins d'enfants for Single-Cycle Belyi Maps* (Manes et al.) describes the dessins d'enfants for two infinite families of dynamical Belyi maps, completing a correspondence given by Riemann's existence theorem.

The last two papers in this volume are in the areas of algebraic number theory, *Multiplicative Order and Frobenius Symbol for the Reductions of Number Fields* (Perucca) and, analytic number theory, *Quantum Modular Forms and Singular Combinatorial Series with Distinct Roots of Unity* (Folsom et al.); the former studies the density of a set of primes of a number field which is defined by some conditions concerning the reductions of algebraic numbers, and the latter establishes the quantum modularity of the $(n + 1)$-variable combinatorial rank generating function for n-marked Durfee symbols.

Workshop Project Titles

WIN4 was a working conference, with several hours each day devoted to research in project groups.

- Apollonian circle packings
  Group members: Holley Friedlander, Elena Fuchs, Piper H, Catherine Hsu, Damaris Schindler, Katherine Stange

- Arithmetic dynamics and Galois representations
  Group members: Jamie Juul, Holly Krieger, Nicole Looper, Michelle Manes, Bianca Thompson, Laura Walton

- Chabauty-Coleman experiments on genus three hyperelliptic curves
  Group members: Jennifer S. Balakrishnan, Francesca Bianchi, Victoria Cantoral-Farfán, Mirela Çiperiani, Anastassia Etropolski

- Computational aspects of supersingular elliptic curves
  Group members: Efrat Bank, Catalina Camacho, Kirsten Eisenträger, Jennifer Park

- Horizontal distribution questions for elliptic curves over $\mathbb{Q}$
  Group members: Chantal David, Ayla Gafni, Amita Malik, Lillian Pierce, Neha Prabhu, Caroline Turnage-Butterbaugh

- Newton polygons of cyclic covers of the projective line
  Group members: Wanlin Li, Elena Mantovan, Rachel Pries, Yunqing Tang

- Quantum modular forms and singular combinatorial series
  Group members: Amanda Folsom, Min-Joo Jang, Sam Kimport, Holly Swisher

- Ramanujan graphs in Cryptography
  Group members: Anamaria Costache, Brooke Feigon, Kristin Lauter, Maike Massierer, Anna Puskás

- Torsion structures on elliptic curves
  Group members: Abbey Bourdon, Özlem Ejder, Yuan Liu, Frances Odumodu, Bianca Viray

Participants and Affiliations at the Time of the Workshop

Jennifer S. Balakrishnan, Boston University, USA
Efrat Bank, University of Michigan, USA
Francesca Bianchi, University of Oxford, UK
Abbey Bourdon, University of Georgia, USA
Ana Catalina Camacho Navarro, Colorado State University, USA
Victoria Cantoral Farfán, ICTP, Italy

Mirela Çiperiani, The University of Texas at Austin, USA
Anamaria Costache, University of Bristol, UK
Chantal David, Concordia University, Canada
Kirsten Eisenträger, The Pennsylvania State University, USA
Özlem Ejder, University of Southern California, USA
Anastassia Etropolski, Rice University, USA
Brooke Feigon, The City College of New York (CUNY), USA
Amanda Folsom, Amherst College, USA
Holley Friedlander, Dickinson College, USA
Elena Fuchs, University of California, Davis, USA
Ayla Gafni, University of Rochester, USA
Piper H, University of Hawai'i at Mānoa, USA
Catherine Hsu, University of Oregon, USA
Min-Joo Jang, University of Cologne, Germany
Jamie Juul, Amherst College, USA
Sam Kimport, Stanford University, USA
Holly Krieger, University of Cambridge, UK
Kristin Lauter, Microsoft Research, USA
Wanlin Li, University of Wisconsin-Madison, USA
Yuan Liu, University of Wisconsin-Madison, USA
Nicole Looper, Northwestern University, USA
Amita Malik, University of Illinois at Urbana-Champaign, USA
Michelle Manes, University of Hawai'i at Mānoa, USA
Elena Mantovan, California Institute of Technology, USA
Maike Massierer, University of New South Wales, Sydney, Australia
Frances Odumodu, Université Bordeaux, France
Jennifer Park, University of Michigan, USA
Lillian Pierce, Duke University, USA
Neha Prabhu, Indian Institute of Science Education and Research-Pune, India
Rachel Pries, Colorado State University, USA
Anna Puskás, University of Alberta, Canada
Damaris Schindler, Utrecht University, The Netherlands
Katherine Stange, University of Colorado Boulder, USA
Holly Swisher, Oregon State University, USA
Yunqing Tang, IAS/Princeton University, USA
Bianca Thompson, Harvey Mudd College, USA
Caroline Turnage-Butterbaugh, Duke University, USA
Bianca Viray, University of Washington, USA
Laura Walton, Brown University, USA

Workshop Website

https://www.birs.ca/events/2017/5-day-workshops/17w5083

Boston, MA, USA   Jennifer S. Balakrishnan
Amherst, MA, USA   Amanda Folsom
Montréal, QC, Canada   Matilde Lalín
Honolulu, HI, USA   Michelle Manes
March 2019

Acknowledgments

We are grateful to the following sponsoring organizations for their support of the workshop and this volume:

- Banff International Research Station
- National Science Foundation (DMS 1712938)
- Clay Mathematics Institute
- Microsoft Research
- The Number Theory Foundation
- Pacific Institute for the Mathematical Sciences
- Association for Women in Mathematics and the AWM ADVANCE Grant (NSF HRD 1500481)

We would like to thank the referees whose careful and dedicated work have been crucial in assuring the quality of this publication.

Contents

List of Contributors

Jennifer S. Balakrishnan Department of Mathematics and Statistics, Boston University, 111 Cummington Mall, Boston, MA 02215, USA, e-mail: jbala@bu.edu

Efrat Bank Department of Mathematics, University of Michigan, Ann Arbor, MI, USA, e-mail: ebank@umich.edu

Francesca Bianchi Mathematical Institute, University of Oxford, Andrew Wiles Building, Radcliffe Observatory Quarter, Woodstock Road, Oxford OX2 6GG, UK, e-mail: francesca.bianchi@maths.ox.ac.uk

Catalina Camacho-Navarro Department of Mathematics, Colorado State University, Fort Collins, CO 80523, USA, e-mail: camacho@math.colostate.edu

Victoria Cantoral-Farfán The Abdus Salam International Center for Theoretical Physics, Mathematics Section, 11 Strada Costiera, 34151 Trieste, Italy, e-mail: vcantora@ictp.it

Mirela Çiperiani Department of Mathematics, The University of Texas at Austin, 1 University Station, C1200, Austin, TX 78712, USA, e-mail: mirela@math.utexas.edu

Anamaria Costache Department of Computer Science, University of Bristol, Bristol, UK, e-mail: anamaria.costache@bristol.ac.uk

Kirsten Eisenträger Department of Mathematics, The Pennsylvania State University, University Park, PA 16802, USA, e-mail: eisentra@math.psu.edu

Anastassia Etropolski Department of Mathematics, Rice University MS 136, Houston, TX 77251, USA, e-mail: aetropolski@rice.edu

Brooke Feigon Department of Mathematics, The City College of New York, CUNY, NAC 8/133, New York, NY 10031, USA, e-mail: bfeigon@ccny.cuny.edu

Amanda Folsom Department of Mathematics and Statistics, Amherst College, Amherst, MA 01002, USA, e-mail: afolsom@amherst.edu

Min-Joo Jang Department of Mathematics, The University of Hong Kong, Room 318, Run Run Shaw Building, Pokfulam, Hong Kong, e-mail: min-joo.jang@hku. hk

Jamie Juul Department of Mathematics, The University of British Columbia, 1984 Mathematics Road, Vancouver, BC V6T 1Z2, Canada, e-mail: jamie.l.rahr@gmail. com

Sam Kimport Department of Mathematics, Stanford University, 450 Serra Mall, Building 380, Stanford, CA 94305, USA, e-mail: skimport@stanford.edu

Holly Krieger Department of Pure Mathematics and Mathematical Statistics, University of Cambridge, Wilberforce Road, Cambridge CB3 0WB, UK, e-mail: hkrieger@dpmms.cam.ac.uk

Kristin Lauter Microsoft Research, One Microsoft Way, Redmond, WA 98052, USA, e-mail: klauter@microsoft.com

Odile Lecacheux Sorbonne Université, Institut de Mathématiques de Jussieu-Paris Rive Gauche, 4 Place Jussieu, 75252 Paris Cedex 05, France, e-mail: odile. lecacheux@imj-prg.fr

Wanlin Li Department of Mathematics, University of Wisconsin, Madison, WI 53706, USA, e-mail: wanlin@math.wisc.edu

Nicole Looper Department of Pure Mathematics and Mathematical Statistics, University of Cambridge, Wilberforce Road, Cambridge CB3 0WB, UK, e-mail: nl393@cam.ac.uk

Michelle Manes Department of Mathematics, University of Hawai'i at Mānoa, 2565 McCarthy Mall Keller 401A, Honolulu, HI 96822, USA, e-mail: mmanes@math.hawaii.edu

Elena Mantovan Department of Mathematics, California Institute of Technology, Pasadena, CA 91125, USA, e-mail: mantovan@caltech.edu

Maike Massierer School of Mathematics and Statistics, University of New South Wales, Sydney, NSW 2052, Australia

Gabrielle Melamed Department of Mathematics, University of Connecticut, 341 Mansfield Road U1009, Storrs, CT 06269, USA, e-mail: Gabrielle. Melamed@uconn.edu

Travis Morrison Institute for Quantum Computing, The University of Waterloo, Waterloo, ON, Canada, e-mail: travis.morrison@uwaterloo.ca

Jennifer Park Department of Mathematics, University of Michigan, Ann Arbor, MI, USA, e-mail: jmypark@umich.edu

Antonella Perucca University of Luxembourg, Mathematics Research Unit, 6, avenue de la Fonte L-4364, Esch-sur-Alzette, Luxembourg, e-mail: antonella. perucca@uni.lu

Rachel Pries Department of Mathematics, Colorado State University, Fort Collins, CO 80523, USA, e-mail: pries@math.colostate.edu

Anna Puskás Department of Mathematics & Statistics, University of Massachusetts, Amherst, MA 01003, USA, e-mail: anna.puskas@ipmu.jp

Holly Swisher Department of Mathematics, Oregon State University, Kidder Hall 368, Corvallis, OR 97331, USA, e-mail: swisherh@math.oregonstate.edu

Yunqing Tang Department of Mathematics, Princeton University, Princeton, NJ 08540, USA, e-mail: yunqingt@math.princeton.edu

Bianca Thompson Westminster College, Salt Lake City, UT 84105, USA, e-mail: bthompson@westminstercollege.edu

Bella Tobin Department of Mathematics, University of Hawai'i at Mānoa, 2565 McCarthy Mall Keller 401A, Honolulu, HI 96822, USA, e-mail: tobin@math.hawaii.edu; bellatobin@gmail.com

Laura Walton Mathematics Department, Brown University, Providence, RI 02912, USA, e-mail: laura@math.brown.edu

Ramanujan Graphs in Cryptography

Anamaria Costache, Brooke Feigon, Kristin Lauter, Maike Massierer,
and Anna Puskás

Abstract In this paper we study the security of a proposal for Post-Quantum Cryptography from both a number theoretic and cryptographic perspective. Charles–Goren–Lauter in 2006 proposed two hash functions based on the hardness of finding paths in Ramanujan graphs. One is based on Lubotzky–Phillips–Sarnak (LPS) graphs and the other one is based on Supersingular Isogeny Graphs. A 2008 paper by Petit–Lauter–Quisquater breaks the hash function based on LPS graphs. On the Supersingular Isogeny Graphs proposal, recent work has continued to build cryptographic applications on the hardness of finding isogenies between supersingular elliptic curves. A 2011 paper by De Feo–Jao–Plût proposed a cryptographic system based on Supersingular Isogeny Diffie–Hellman as well as a set of five hard problems. In this paper we show that the security of the SIDH proposal relies on

Brooke Feigon was partially supported by National Security Agency grant H98230-16-1-0017 and PSC-CUNY.
Maike Massierer was partially supported by Australian Research Council grant DP150101689.

A. Costache
Department of Computer Science, University of Bristol, Bristol, UK
e-mail: anamaria.costache@bristol.ac.uk

B. Feigon (✉)
Department of Mathematics, The City College of New York, CUNY, NAC 8/133, New York, NY 10031, USA
e-mail: bfeigon@ccny.cuny.edu

K. Lauter
Microsoft Research, One Microsoft Way, Redmond, WA 98052, USA
e-mail: klauter@microsoft.com

M. Massierer
School of Mathematics and Statistics, University of New South Wales, Sydney, NSW 2052, Australia

A. Puskás
Department of Mathematics & Statistics, University of Massachusetts, Amherst, MA 01003, USA
e-mail: anna.puskas@ipmu.jp

the hardness of the SSIG path-finding problem introduced in Charles et al. (2009). In addition, similarities between the number theoretic ingredients in the LPS and Pizer constructions suggest that the hardness of the path-finding problem in the two graphs may be linked. By viewing both graphs from a number theoretic perspective, we identify the similarities and differences between the Pizer and LPS graphs.

Keywords Post-Quantum Cryptography · Supersingular isogeny graphs · Ramanujan graphs

2010 Mathematics Subject Classification Primary: 14G50, 11F70; Secondary: 05C75, 11R52

1 Introduction

Supersingular Isogeny Graphs were proposed for use in cryptography in 2006 by Charles, Goren, and Lauter [3]. Supersingular isogeny graphs are examples of Ramanujan graphs, i.e., optimal expander graphs. This means that relatively *short* walks on the graph approximate the uniform distribution, i.e., walks of length approximately equal to the logarithm of the graph size. Walks on expander graphs are often used as a good source of randomness in computer science, and the reason for using *Ramanujan* graphs is to keep the path length short. But the reason these graphs are important for cryptography is that *finding paths* in these graphs, i.e., *routing*, is hard: there are no known subexponential algorithms to solve this problem, either classically or on a quantum computer. For this reason, systems based on the hardness of problems on Supersingular Isogeny Graphs are currently under consideration for standardization in the NIST Post-Quantum Cryptography (PQC) Competition [21].

Charles et al. [3] proposed a general construction for cryptographic hash functions based on the hardness of inverting a walk on a graph. The path-finding problem is the following: given fixed starting and ending vertices representing the start and end points of a walk on the graph of a fixed length, find a path between them. A hash function can be defined by using the input to the function as directions for walking around the graph: the output is the label for the ending vertex of the walk. Finding collisions for the hash function is equivalent to finding cycles in the graph, and finding preimages is equivalent to path-finding in the graph. Backtracking is not allowed in the walks by definition, to avoid trivial collisions.

In [3], two concrete examples of families of optimal expander graphs (Ramanujan graphs) were proposed, the so-called Lubotzky–Phillips–Sarnak (LPS) graphs [14], and the Supersingular Isogeny Graphs (Pizer) [20], where the path-finding problem was supposed to be hard. Both graphs were proposed and presented at the 2005 and 2006 NIST Hash Function workshops, but the LPS hash function was quickly attacked and broken in two papers in 2008, a collision attack [24] and a preimage attack [17]. The preimage attack gives an algorithm to efficiently find paths in LPS graphs, a problem which had been open for several decades. The PLQ path-finding algorithm uses the explicit description of the graph as a Cayley graph in $\mathrm{PSL}_2(\mathbb{F}_p)$, where vertices are 2×2 matrices with entries in $\mathbb{F}_p$ satisfying certain properties. Given the swift discovery of attacks on the LPS path-finding problem, it is natural to investigate whether this approach is relevant to the path-finding problem in Supersingular Isogeny (Pizer) Graphs.

In 2011, De Feo–Jao–Plût [8] devised a cryptographic system based on supersingular isogeny graphs, proposing a Diffie–Hellman protocol as well as a set of five hard problems related to the security of the protocol. It is natural to ask what is the relation between the problems stated in [8] and the path-finding problem on Supersingular Isogeny Graphs proposed in [3].

In this paper we explore these two questions related to the security of cryptosystems based on these Ramanujan graphs. In Part 1 of the paper, we study the relation between the hard problems proposed by De Feo–Jao–Plût and the hardness of the Supersingular Isogeny Graph problem which is the foundation for the CGL hash function. In Part 2 of the paper, we study the relation between the Pizer and LPS graphs by viewing both from a number theoretic perspective.

In particular, in Part 1 of the paper, we clearly explain how the security of the Key-Exchange protocol relies on the hardness of the path-finding problem in SSIG, proving a reduction (Theorem 3.2) between the Supersingular Isogeny Diffie Hellmann (SIDH) Problem and the path-finding problem in SSIG. Although this fact and this theorem may be clear to the experts (see, for example, the comment in the introduction to a recent paper on this topic [1]), this reduction between the hard problems is not written anywhere in the literature. Furthermore, the Key-Exchange (SIDH) paper [8] states 5 hard problems, including (SSCDH), with relations proved between some but not all of them, and mentions the paper [3] only in passing (on page 17), with no clear statement of the relationship to the overarching hard problem of path-finding in SSIG.

Our Theorem 3.2 clearly shows the fact that the security of the proposed post-quantum key-exchange relies on the hardness of the path-finding problem in SSIG stated in [3]. Theorem 4.9 counts the chains of isogenies of fixed length. Its proof relies on elementary group theory results and facts about isogenies, proved in Section 4.

In Part 2 of the paper, we examine the LPS and Pizer graphs from a number theoretic perspective with the aim of highlighting the similarities and differences between the constructions.

Both the LPS and Pizer graphs considered in [3] can be thought of as graphs on

$$\Gamma \backslash \mathrm{PGL}_2(\mathbb{Q}_l) / \mathrm{PGL}_2(\mathbb{Z}_l), \tag{1}$$

where Γ is a discrete cocompact subgroup, where Γ is obtained from a quaternion algebra B. We show how different input choices for the construction lead to different graphs. In the LPS construction one may vary Γ to get an infinite family of Ramanujan graphs. In the Pizer construction one may vary B to get an infinite family. In the LPS case, we always work in the Hamiltonian quaternion algebra. For this particular choice of algebra we can rewrite the graph as a Cayley graph. This explicit description is key for breaking the LPS hash function. For the Pizer graphs we do not have such a description. On the Pizer side the graphs may, via Strong Approximation, be viewed as graphs on adèlic double cosets which are in turn the class group of an order of B that is related to the cocompact subgroup Γ. From here one obtains an isomorphism with supersingular isogeny graphs. For LPS graphs the local double cosets are also isomorphic to adèlic double cosets, but in this case the corresponding set of adèlic double cosets is smaller relative to the quaternion algebra and we do not have the same chain of isomorphisms.

Part 2 has the following outline. Section 6 follows [15] and presents the construction of LPS graphs from three different perspectives: as a Cayley graph, in terms of local double cosets, and, to connect these two, as a quotient of an infinite tree. The edges of the LPS graph are explicit in both the Cayley and local double coset presentation. In Section 6.4 we give an explicit bijection between the natural parameterizations of the edges at a fixed vertex. Section 7 is about Strong Approximation, the main tool connecting the local and adelic double cosets for both LPS and Pizer graphs. Section 8 follows [20] and summarizes Pizer's construction. The different input choices for LPS and Pizer constructions impose different restrictions on the parameters of the graph, such as the degree. 6-regular graphs exist in both families. In Section 8.2 we give a set of congruence conditions for the parameters of the Pizer construction that produce a 6-regular graph. In Section 9 we summarize the similarities and differences between the two constructions.

1.1 Acknowledgments

This project was initiated at the Women in Numbers 4 (WIN4) workshop at the Banff International Research Station in August, 2017. The authors would like to thank BIRS and the WIN4 organizers. In addition, the authors would like to thank the Clay Mathematics Institute, PIMS, Microsoft Research, the Number Theory Foundation, and the NSF-HRD 1500481—AWM ADVANCE grant for supporting the workshop. We thank John Voight, Scott Harper, and Steven Galbraith for helpful conversations, and the anonymous referees for many helpful suggestions and edits.

Part 1: Cryptographic Applications of Supersingular Isogeny Graphs

In this section we investigate the security of the [8] key-exchange protocol. We show a reduction to the path-finding problem in supersingular isogeny graphs stated in [3]. The hardness of this problem is the basis for the CGL cryptographic hash function, and we show here that if this problem is not hard, then the key exchange presented in [8] is not secure.

We begin by recalling some basic facts about isogenies of elliptic curves and the key-exchange construction. Then, we give a reduction between two hardness assumptions. This reduction is based on a correspondence between a path representing the composition of m isogenies of degree ℓ and an isogeny of degree ℓ^m.

2 Preliminaries

We start by recalling some basic and well-known results about isogenies. They can all be found in [23]. We try to be as concrete and constructive as possible, since we would like to use these facts to do computations.

An elliptic curve is a curve of genus one with a specific base point $\mathcal{O}$. This latter can be used to define a group law. We will not go into the details of this, see, for example, [23]. If E is an elliptic curve defined over a field K and $\mathrm{char}(\bar{K}) \neq 2, 3$, we can write the equation of E as

$$E : y^2 = x^3 + a \cdot x + b,$$

where $a, b \in K$. Two important quantities related to an elliptic curve are its discriminant Δ and its j-invariant, denoted by j. They are defined as follows:

$$\Delta = 16 \cdot (4 \cdot a^3 + 27 \cdot b^2) \quad \text{and} \quad j = -1728 \cdot \frac{a^3}{\Delta}.$$

Two elliptic curves are isomorphic over $\bar{K}$ if and only if they have the same j-invariant.

Definition 2.1. *Let E_0 and E_1 be two elliptic curves. An isogeny from E_0 to E_1 is a surjective morphism*

$$\phi : E_0 \to E_1,$$

which is a group homomorphism.

An example of an isogeny is the multiplication-by-m map $[m]$,

$$[m] : E \to E$$

$$P \mapsto m \cdot P.$$

The degree of an isogeny is defined as the degree of the finite extension $\bar{K}(E_0)/\phi^*(\bar{K}(E_1))$, where $\bar{K}(*)$ is the function field of the curve, and ϕ^* is the map of function fields induced by the isogeny ϕ. By convention, we set

$$\deg([0]) = 0.$$

The degree map is multiplicative under composition of isogenies:

$$\deg(\phi \circ \psi) = \deg(\phi) \cdot \deg(\psi)$$

for all chains $E_0 \xrightarrow{\phi} E_1 \xrightarrow{\psi} E_2$, and for an integer $m > 0$, the multiplication-by-m map has degree m^2.

Theorem 2.2 ([23]). *Let $E_0 \to E_1$ be an isogeny of degree m. Then, there exists a unique isogeny*

$$\hat{\phi} : E_1 \to E_0$$

such that $\hat{\phi} \circ \phi = [m]$ on E_0, and $\phi \circ \hat{\phi} = [m]$ on E_1. We call $\hat{\phi}$ the dual isogeny to ϕ. We also have that

$$\deg(\hat{\phi}) = \deg(\phi).$$

For an isogeny ϕ, we say ϕ is separable if the field extension $\bar{K}(E_0)/\phi^*(\bar{K}(E_1))$ is separable. We then have the following lemma.

Lemma 2.3. *Let $\phi : E_0 \to E_1$ be a separable isogeny. Then*

$$\deg(\phi) = \#\ker(\phi).$$

In this paper, we only consider separable isogenies and frequently use this convenient fact. From the above, it follows that a point P of order m defines an isogeny ϕ of degree m,

$$\phi : E \to E/\langle P \rangle.$$

We will refer to such an isogeny as a cyclic isogeny (meaning that its kernel is a cyclic subgroup of E). For ℓ prime, we also say that two curves E_0 and E_1 are ℓ-isogenous if there exists an isogeny $\phi : E_0 \to E_1$ of degree ℓ.

We define $E[m]$, the m-torsion subgroup of E, to be the kernel of the multiplication-by-m map. If $\text{char}(K) > 0$ and $m \geq 2$ is an integer coprime to $\text{char}(K)$, or if $\text{char}(K) = 0$, then the points of $E[m]$ are

$$E[m] = \{P \in E(\bar{K}) : m \cdot P = \mathcal{O}\} \cong \mathbb{Z}/m\mathbb{Z} \times \mathbb{Z}/m\mathbb{Z}.$$

If an elliptic curve E is defined over a field of characteristic $p > 0$ and its endomorphism ring over $\bar{K}$ is an order in a quaternion algebra, we say that E is supersingular. Every isomorphism class over $\bar{K}$ of supersingular elliptic curves in characteristic p has a representative defined over $\mathbb{F}_{p^2}$, thus we will often let $K = \mathbb{F}_{p^2}$ (for some fixed prime p).

We mentioned above that an ℓ-torsion point P induces an isogeny of degree ℓ. More generally, a finite subgroup G of E generates a unique isogeny of degree #G, up to automorphism.

Supersingular isogeny graphs were introduced into cryptography in [3]. To define a supersingular isogeny graph, fix a finite field K of characteristic p, a supersingular elliptic curve E over K, and a prime $\ell \neq p$. Then the corresponding isogeny graph is constructed as follows. The vertices are the $\bar{K}$-isomorphism classes of elliptic curves which are $\bar{K}$-isogenous to E. Each vertex is labeled with the j-invariant of the curve. The edges of the graph correspond to the ℓ-isogenies between the elliptic curves. As the vertices are isomorphism classes of elliptic curves, isogenies that differ by composition with an automorphism of the image are identified as edges of the graph. That is, if E_0, E_1 are $\bar{K}$-isogenous elliptic curves, $\phi : E_0 \rightarrow E_1$ is an ℓ-isogeny and $\epsilon \in \mathrm{Aut}(E_1)$ is an automorphism, then ϕ and $\epsilon \circ \phi$ are identified and correspond to the same edge of the graph.

If $p \equiv 1 \mod 12$, we can uniquely identify an isogeny with its dual to make it an undirected graph. It is a multigraph in the sense that there can be multiple edges if no extra conditions are imposed on p. Three important properties of these graphs follow from deep theorems in number theory:

1. The graph is connected for any $\ell \neq p$ (special case of [4, Theorem 4.1]).
2. A supersingular isogeny graph has roughly $p/12$ vertices [23, Theorem 4.1].
3. Supersingular isogeny graphs are optimal expander graphs, in particular they are Ramanujan (special case of [4, Theorem 4.2]).

Remark 2.4. In order to avoid trivial collisions in cryptographic hash functions based on isogeny graphs, it is best if the graph has no short cycles. Charles, Goren, and Lauter show in [3] how to ensure that isogeny graphs do not have short cycles by carefully choosing the finite field one works over. For example, they compute that a 2-isogeny graph does not have double edges (i.e., cycles of length 2) when working over $\mathbb{F}_p$ with $p \equiv 1 \mod 420$. Similarly, we computed that a 3-isogeny graph does not have double edges for $p \equiv 1 \mod 9240$. Given that $420 = 2^2 \cdot 3 \cdot 5 \cdot 7$ and $9240 = 2^3 \cdot 3 \cdot 5 \cdot 7 \cdot 11$, we conclude that neither the 2-isogeny graph nor the 3-isogeny graph has double edges for $p \equiv 1 \mod 9240$.

For our experiments (described in Section 4), we were interested in studying short walks, for example, of length 4, in a setting relevant to the Key-Exchange protocol described below. The smallest prime p with the property $p \equiv 1 \mod 9240$ that also satisfies $2^4 \cdot 3^4 \mid p - 1$ is

$$p = 2^4 \cdot 3^4 \cdot 5 \cdot 7 \cdot 11 + 1.$$

3 The [8] Key-Exchange

Let E be a supersingular elliptic curve defined over $\mathbb{F}_{p^2}$, where $p = \ell_A^n \cdot \ell_B^m \pm 1$, ℓ_A and ℓ_B are primes, and $n \approx m$ are approximately equal. We have players A (for Alice) and B (for Bob), representing the two parties who wish to engage in a key-exchange protocol with the goal of establishing a shared secret key by communicating via a (possibly) insecure channel. The two players A and B generate their public parameters by each picking two points P_A, Q_A such that $\langle P_A, Q_A \rangle = E[\ell_A^n]$ (for A), and two points P_B, Q_B such that $\langle P_B, Q_B \rangle = E[\ell_B^m]$ (for B).

Player A then secretly picks two random integers $0 \leq m_A, n_A < \ell_A^n$. These two integers (and the isogeny they generate) will be player A's secret parameters. A then computes the isogeny ϕ_A

$$E \xrightarrow{\phi_A} E_A := E/\langle [m_A]P_A + [n_A]Q_A \rangle.$$

Player B proceeds in a similar fashion and secretly picks $0 \leq m_B, n_B < \ell_B^m$. Player B then generates the (secret) isogeny

$$E \xrightarrow{\phi_B} E_B := E/\langle [m_B]P_B + [n_B]Q_B \rangle.$$

So far, A and B have constructed the following diagram:

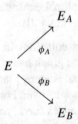

To complete the diamond, we proceed to the exchange part of the protocol. Player A computes the points $\phi_A(P_B)$ and $\phi_A(Q_B)$ and sends $\{\phi_A(P_B), \phi_A(Q_B), E_A\}$ along to player B. Similarly, player B computes and sends $\{\phi_B(P_A), \phi_B(Q_A), E_B\}$ to player A. Both players now have enough information to construct the following diagram:

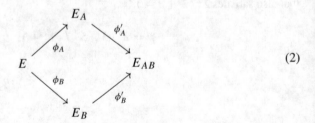

$$(2)$$

where

$$E_{AB} \cong E/\langle [m_A]P_A + [n_A]Q_A, [m_B]P_B + [n_B]Q_B \rangle.$$

Player A can use the knowledge of the secret information m_A and n_A to compute the isogeny ϕ_B', by quotienting E_B by $\langle [m_A]\phi_B(P_A)+[n_A]\phi_B(Q_A) \rangle$ to obtain E_{AB}. Player B can use the knowledge of the secret information m_B and n_B to compute the isogeny ϕ_A', by quotienting E_A by $\langle [m_B]\phi_A(P_B)+[n_B]\phi_A(Q_B) \rangle$ to obtain E_{AB}. A separable isogeny is determined by its kernel, and so both ways of going around the diagram from E result in computing the same elliptic curve E_{AB}.

The players then use the j-invariant of the curve E_{AB} as a shared secret.

Remark 3.1. Given a list of points specifying a kernel, one can explicitly compute the associated isogeny using Vélu's formulas [25]. In principle, this is how the two parties engaging in the key exchange above can compute ϕ_A, ϕ_B, ϕ_A', ϕ_B' [25]. However, in practice for cryptographic size subgroups, this would be impossible, and thus a different approach is taken, based on breaking the isogenies into n (resp. m) steps, each of degree ℓ_A (resp. ℓ_B). This equivalence will be explained below.

3.1 Hardness Assumptions

The security of the key-exchange protocol is based on the following hardness assumption, which was introduced in [8] and called the Supersingular Computational Diffie–Hellman (SSCDH) problem.

Problem 1 (Supersingular Computational Diffie–Hellman (SSCDH)). *Let p, ℓ_A, ℓ_B, n, m, E, E_A, E_B, E_{AB}, P_A, Q_A, P_B, Q_B be as above.*

Let ϕ_A be an isogeny from E to E_A whose kernel is equal to $\langle [m_A]P_A + [n_A]Q_A \rangle$, and let ϕ_B be an isogeny from E to E_B whose kernel is equal to $\langle [m_B]P_B + [n_B]Q_B \rangle$, where m_A,n_A (respectively m_B,n_B) are integers chosen at random between 0 and ℓ_A^m (respectively ℓ_B^n), and not both divisible by ℓ_A (resp. ℓ_B).

Given the curves E_A, E_B and the points $\phi_A(P_B)$, $\phi_A(Q_B)$, $\phi_B(P_A)$, $\phi_B(Q_A)$, find the j-invariant of

$$E_{AB} \cong E/\langle [m_A]P_A + [n_A]Q_A, [m_B]P_B + [n_B]Q_B \rangle;$$

see diagram (2).

In [3], a cryptographic hash function was defined:

$$h : \{0, 1\}^r \rightarrow \{0, 1\}^s$$

based on the Supersingular Isogeny Graph (SSIG) for a fixed prime p of cryptographic size, and a fixed small prime $\ell \neq p$. The hash function processes the input string in blocks which are used as directions for walking around the graph starting from a given fixed vertex. The output of the hash function is the j-invariant of an elliptic curve over $\mathbb{F}_{p^2}$ which requires $2\log(p)$ bits to represent, so $m = 2\lceil\log(p)\rceil$. For the security of the hash function, it is necessary to avoid the generic *birthday attack*. This attack runs in time proportional to the square root of the size of the graph, which is the *Eichler class number*, roughly $\lfloor p/12 \rfloor$. So in practice, we must pick p so that $\log(p) \approx 256$.

The integer r is the length of the bit string input to the hash function. If $\ell = 2$, which is the easiest case to implement and a common choice, then r is precisely the number of steps taken on the walk in the graph, since the graph is 3-regular, with no backtracking allowed, so the input is processed bit-by-bit. In order to assure that the walk reaches a sufficiently random vertex in the graph, the number of steps should be roughly $\log(p) \approx 256$. A CGL-hash function is thus specified by giving the primes p, ℓ, the starting point of the walk, and the integers $r \approx 256$, s. (Extra congruence conditions were imposed on p to make it an undirected graph with no small cycles.)

The hard problems stated in [3] corresponded to the important security properties of *collision* and *preimage resistance* for this hash function. For preimage resistance, the problem [3, Problem 3] stated was: given p, ℓ, $r > 0$, and two supersingular j-invariants modulo p, to find a path of length r between them:

Problem 2 (Path-Finding [3]). *Let p and ℓ be distinct prime numbers, $r > 0$, and E_0 and E_1 two supersingular elliptic curves over $\mathbb{F}_{p^2}$. Find a path of length r in the ℓ-isogeny graph corresponding to a composition of r ℓ-isogenies leading from E_0 to E_1 (i.e., an isogeny of degree ℓ^r from E_0 to E_1).*

It is worth noting that, to break the preimage resistance of the specified hash function, you must find a path of exactly length r, and this is analogous to the situation for breaking the security of the key-exchange protocol. However, the problem of finding *any* path between two given vertices in the SSIG graphs is also still open. For the LPS graphs, the algorithm presented in [17] did not find a path of a specific given length, but it was still considered to be a "break" of the hash function.

Furthermore, the diameter of these graphs, both LPS and SSIG graphs, has been extensively studied. It is known that the diameter of the graphs is roughly $\log(p)$ (it is $c\log(p)$, where c is a constant between 1 and 2, (see, for example, [22])). That means that if r is greater than $c\log(p)$, then given two vertices, it is likely that a path of length r between them may exist. The fact that walks of length greater than $c\log(p)$ approximate the uniform distribution very closely means that you are not likely to miss any significant fraction of the vertices with paths of that length, because that would constitute a bias. Also, if $r \gg \log(p)$, then there may be many paths of length r. However, if r is much less than $\log(p)$, such as $\frac{1}{2}\log(p)$, there may be *no path* of such a short length between two given vertices. See [13] for a discussion of the "sharp cutoff" property of Ramanujan graphs.

But in the cryptographic applications, given an instance of the key-exchange protocol to be attacked, we *know* that there exists a path of length n between E

and E_A, and the hard problem is to find it. The setup for the key exchange requires $p = \ell_A^n \ell_B^m \pm 1$, where n and m are roughly the same size, and ℓ_A and ℓ_B are very small, such as $\ell_A = 2$ and $\ell_B = 3$. It follows that n and m are both approximately half the diameter of the graph (which is roughly $\log(p)$). So it is unlikely to find paths of length n or m between two random vertices. If a path of length n exists and Algorithm A finds a path, then it is very likely to be the one which was constructed in the key-exchange. If not, then Algorithm A can be repeated any constant number of times. So we have the following reduction:

Theorem 3.2. *Assume as for the Key-Exchange setup that* $p = \ell_A^n \cdot \ell_B^m + 1$ *is a prime of cryptographic size, i.e.,* $\log(p) \geq 256$, ℓ_A *and* ℓ_B *are small primes, such as* $\ell_A = 2$ *and* $\ell_B = 3$, *and* $n \approx m$ *are approximately equal. Given an algorithm to solve Problem 2 (Path-Finding), it can be used to solve Problem 1 (Key-Exchange) with overwhelming probability. The failure probability is roughly*

$$\frac{\ell_A^n + \ell_A^{n-1}}{p} \approx \frac{\sqrt{p}}{p}.$$

Proof. Given an algorithm (Algorithm A) to solve Problem 2, we can use this to solve Problem 1 as follows. Given E and E_A, use Algorithm A to find the path of length n between these two vertices in the ℓ_A-isogeny graph. Now use Lemma 4.4 below to produce a point R_A which generates the ℓ_A^n-isogeny between E and E_A. Repeat this to produce the point R_B which generates the ℓ_B^m-isogeny between E and E_B in the ℓ_B-isogeny graph. Because the subgroups generated by R_A and R_B have smooth order, it is easy to write R_A in the form $[m_A]P_A + [n_A]Q_A$ and R_B in the form $[m_B]P_B + [n_B]Q_B$. Using the knowledge of m_A, n_A, m_B, n_B, we can construct E_{AB} and recover the j-invariant of E_{AB}, allowing us to solve Problem 1.

The reason for the qualification "with overwhelming probability" in the statement of the theorem is that it is possible that there are multiple paths of the same length between two vertices in the graph. If there are multiple paths of length n (or m) between the two vertices, it suffices to repeat Algorithm A to find another path. This approach is sufficient to break the Key-Exchange if there are only a small number of paths to try. As explained above, with overwhelming probability, there are *no* other paths of length n (or m) in the Key-Exchange setting.

In the SSIG corresponding to (p, ℓ_A), the vertices E and E_A are a distance of n apart. Starting from the vertex E and considering all paths of length n, the number of possible endpoints is at most $\ell_A^n + \ell_A^{n-1}$ (see Corollary 4.8 below). Considering that the number of vertices in the graph is roughly $\lfloor p/12 \rfloor$, then the probability that a given vertex such as E_A will be the endpoint of one of the walks of length n is roughly

$$\frac{\ell_A^n + \ell_A^{n-1}}{p} \approx \frac{\sqrt{p}}{p} \leq 2^{-128}.$$

This estimate does not use the Ramanujan property of the SSIG graphs. While a generic random graph could potentially have a topology which creates a bias towards some subset of the nodes, Ramanujan graphs cannot, as shown in [13, Theorem 3.5]. □

4 Composing Isogenies

Let k be a positive integer. Every separable k-isogeny $\phi : E_0 \to E_1$ is determined by its kernel up to composition with an automorphism of the elliptic curve E_1. Thus the edge corresponding to ϕ is uniquely determined by $\ker(\phi)$ and vice versa. This kernel is a subgroup of the k-torsion $E_0[k]$, and the latter is isomorphic to $\mathbb{Z}/k\mathbb{Z} \times \mathbb{Z}/k\mathbb{Z}$ if k is coprime to the characteristic of the field we are working over.

Hence, fixing a prime ℓ and working over a finite field $\mathbb{F}_q$ which has characteristic different from ℓ, the number of ℓ-isogenies $\phi : E_0 \to E_1$ that correspond to different edges of the graph is equal to the number of subgroups of $\mathbb{Z}/\ell\mathbb{Z} \times \mathbb{Z}/\ell\mathbb{Z}$ of order ℓ. It is well known that this number is equal to $\ell + 1$. In other words, E is ℓ-isogenous to precisely $\ell + 1$ elliptic curves.

However, some of these ℓ-isogenous curves may be isomorphic. Therefore, in the isogeny graph (where nodes represent isomorphism classes of curves), E has degree $\ell + 1$ and may have $\ell + 1$ neighbors or fewer.

Using Vélu's formulas, the equations for an edge can be computed from its kernel. Hence for computational purposes, it is important to write down this kernel explicitly. This is best done by specifying generators. Let $P, Q \in E_0$ be the generators of $E_0[\ell] \cong \mathbb{Z}/\ell\mathbb{Z} \times \mathbb{Z}/\ell\mathbb{Z}$. Then the subgroups of order ℓ are generated by Q and $P + iQ$ for $i = 0, \ldots, \ell - 1$.

We now study isogenies obtained by composition, and isogenies of degree a prime power. It turns out that these correspond to each other under certain conditions. The first condition is that the isogeny is cyclic. Notice that every prime order group is cyclic, therefore all ℓ-isogenies are cyclic (meaning they have cyclic kernel). However, this is not necessarily true for isogenies whose order is not a prime. The second condition is that there is no backtracking, defined as follows:

Definition 4.1. *For a chain of isogenies* $\phi_m \circ \phi_{m-1} \circ \ldots \circ \phi_1$ ($\phi_i : E_{i-1} \to E_i$), *we say that it has no backtracking if* $\phi_{i+1} \neq \epsilon \circ \hat{\phi}_i$ *for all* $i = 1, \ldots, m - 1$ *and any* $\epsilon \in \mathrm{Aut}(E_{i+1})$, *since this corresponds to a walk in the ℓ-isogeny graph without backtracking.*

In the following, we show that chains of ℓ-isogenies of length m without backtracking correspond to cyclic ℓ^m-isogenies. Recall that we are only considering separable isogenies throughout.

Lemma 4.2. *Let ℓ be a prime, and let ϕ be a separable ℓ^m-isogeny with cyclic kernel. Then there exist cyclic ℓ-isogenies $\phi_1, \ldots, \phi_m$ such that $\phi = \phi_m \circ \phi_{m-1} \circ \ldots \circ \phi_1$ without backtracking.*

Proof. Assume that $\phi = E_0 \to E$, and that its kernel is $\langle P_0 \rangle \subseteq E_0$, where P_0 has order ℓ^m. For $i = 1, \ldots, m$, let

$$\phi_i : E_{i-1} \to E_i$$

be an isogeny with kernel $\langle \ell^{m-i} P_{i-1} \rangle$, where $P_i = \phi_i(P_{i-1})$.

We show that ϕ_i is an ℓ-isogeny for $i \in \{1, \ldots, m\}$ by observing that $\ell^{m-i} P_{i-1}$ has order ℓ. The statement is trivial for $i = 1$. For $i \geq 2$, clearly $\ell^{m-i} P_{i-1} = \ell^{m-i} \phi_{i-1}(P_{i-2}) = \phi_{i-1}(\ell^{m-i} P_{i-2}) \neq \mathcal{O}$, since $\ell^{m-i} P_{i-2} \notin \ker \phi_{i-1} = \langle \ell^{m-(i-1)} P_{i-2} \rangle = \{\ell^{m-(i-1)} P_{i-2}, 2\ell^{m-(i-1)} P_{i-2}, \ldots, (\ell - 1)\ell^{m-(i-1)} P_{i-2}\}$. Furthermore, $\ell \cdot \ell^{m-i} P_{i-1} = \ell^{m-(i-1)} \phi_{i-1}(P_{i-2}) = \phi_{i-1}(\ell^{m-(i-1)} P_{i-2}) = \mathcal{O}$, using the definition of $\ker \phi_{i-1}$.

Next, we show by induction that $\phi_i \circ \ldots \circ \phi_1$ has kernel $\langle \ell^{m-i} P_0 \rangle$. Then it follows that $\phi_m \circ \ldots \circ \phi_1$ is the same as ϕ up to an automorphism ϵ of E, since the two have the same kernel. Replacing ϕ_m with $\epsilon \circ \phi_m$ if necessary we have $\phi = \phi_m \circ \phi_{m-1} \circ \ldots \circ \phi_1$. The case $i = 1$ is trivial: $\phi_1 : E_0 \to E_1$ has kernel $\langle \ell^{m-1} P_0 \rangle$ by definition. Now assume the statement is true for $i - 1$. Then, we have $\langle \ell^{m-i} P_0 \rangle \subseteq \ker \phi_i \circ \ldots \circ \phi_1$. Conversely, let $Q \in \ker \phi_i \circ \ldots \circ \phi_1$. Then $\phi_{i-1} \circ \ldots \circ \phi_i(Q) \in \ker \phi_i = \langle \ell^{m-i} P_{i-1} \rangle = \phi_{i-1}(\langle \ell^{m-i} P_{i-2} \rangle) = \ldots = \phi_{i-1} \circ \ldots \circ \phi_1(\langle \ell^{m-i} P_0 \rangle)$ and hence $Q \in \langle \ell^{m-i} P_0 \rangle + \ker \phi_{i-1} \circ \ldots \circ \phi_1 = \langle \ell^{m-i} P_0 \rangle + \langle \ell^{m-(i-1)} P_0 \rangle = \langle \ell^{m-i} P_0 \rangle$.

Finally, we show that there is no backtracking in $\phi_m \circ \ldots \circ \phi_1$. Contrarily, assume that there is an $i \in \{1, \ldots, m-1\}$ and $\epsilon \in \mathrm{Aut}(E_{i+1})$ such that $\phi_{i+1} = \epsilon \circ \hat{\phi}_i$. Then, since $\ker(\phi_{i+1} \circ \phi_i) = \ker(\epsilon \circ \hat{\phi}_i \circ \phi_i) = \ker([\ell])$, we have $\ker(\phi_{i+1} \circ \phi_i \circ \phi_{i-1} \circ \ldots \circ \phi_1) = \ker([\ell] \circ \phi_{i-1} \circ \ldots \circ \phi_1)$. Notice that $[\ell]$ commutes with all ϕ_j, and hence $E_0[\ell] \subseteq \ker(\phi_{i+1} \circ \phi_i \circ \phi_{i-1} \circ \ldots \circ \phi_1) \subseteq \ker(\phi_m \circ \phi_i \circ \phi_{i-1} \circ \ldots \circ \phi_1) = \ker \phi$. Since $E_0[\ell] \cong \mathbb{Z}/\ell\mathbb{Z} \times \mathbb{Z}/\ell\mathbb{Z}$, the kernel of ϕ cannot be cyclic, a contradiction. $\square$

Remark 4.3. It is clear that in the above lemma, if ϕ is defined over a finite field $\mathbb{F}_q$, then all ϕ_i are also defined over this field. Namely, if E_0 is defined over $\mathbb{F}_q$ and the kernel is generated by an $\mathbb{F}_q$-rational point, then by Vélu we obtain $\mathbb{F}_q$-rational formulas for ϕ_1, which means that ϕ_1 is defined over $\mathbb{F}_q$, and so on.

Lemma 4.4. *Let ℓ be a prime, let E_i be elliptic curves for $i = 0, \ldots, m$, and let $\phi_i : E_{i-1} \to E_i$ be ℓ-isogenies for $i = 1, \ldots, m$ such that $\phi_{i+1} \neq \epsilon \circ \hat{\phi}_i$ for $i = 1, \ldots, m-1$ and any $\epsilon \in \mathrm{Aut}(E_{i+1})$ (i.e., there is no backtracking). Then $\phi_m \circ \ldots \circ \phi_1$ is a cyclic ℓ^m-isogeny.*

Proof. The degree of isogenies multiplies when they are composed, see, e.g., [23, Ch. III.4]. Hence we are left with proving that the composition of the isogenies is cyclic.

First note that all ϕ_i are cyclic since they have prime degree, and denote by $P_{i-1} \in E_{i-1}$ the generators of the respective kernels. Let Q_{m-1} be a point on E_{m-1} such that $\ell Q_{m-1} = P_{m-1}$. Notice that such a point always exists over the algebraic closure of the field of definition of the curve. Let $R_{m-2} = \hat{\phi}_{m-1}(Q_{m-1})$, where the hat denotes the dual isogeny. Then $\phi_m \circ \phi_{m-1}(R_{m-2}) = \phi_m \circ \phi_{m-1} \circ \hat{\phi}_{m-1}(Q_{m-1}) =$

$\phi_m \circ [\ell](Q_{m-1}) = \phi_m(\ell Q_{m-1}) = \phi_m(P_{m-1}) = \mathcal{O}$, and hence R_{m-2} is in the kernel of $\phi_m \circ \phi_{m-1}$.

Next we show that R_{m-2} has order ℓ^2, which implies that it generates the kernel of $\phi_m \circ \phi_{m-1}$. Suppose that $\ell R_{m-2} = \mathcal{O}$. Then $\mathcal{O} = \ell R_{m-2} = \ell \hat{\phi}_{m-1}(Q_{m-1}) = \hat{\phi}_{m-1}(P_{m-1})$. Since P_{m-1} has order ℓ, this implies that P_{m-1} generates the kernel of $\hat{\phi}_{m-1}$. However, P_{m-1} also generates the kernel of ϕ_m, so $\epsilon \circ \hat{\phi}_{m-1} = \phi_m$ for some $\epsilon \in \mathrm{Aut}(E_m)$. But this is a contradiction to the assumption of no backtracking.

By iterating this argument, we obtain a point R_0 which generates the kernel of $\phi_m \circ \dots \circ \phi_1$, and hence this isogeny is cyclic. $\square$

Combining Lemmas 4.2 and 4.4, we obtain the following correspondence.

Corollary 4.5. *Let ℓ be a prime and m a positive integer. There is a one-to-one correspondence between cyclic separable ℓ^m-isogenies and chains of separable ℓ-isogenies of length m without backtracking. (Here we do not distinguish between isogenies that differ by composition with an automorphism on the image.)*

Next, we investigate how many such isogenies there are. We start by studying ℓ^m-isogenies. The following group theory result is crucial.

Lemma 4.6. *Let ℓ be a prime and m a positive integer. Then the number of subgroups of $\mathbb{Z}/\ell^m\mathbb{Z} \times \mathbb{Z}/\ell^m\mathbb{Z}$ of order ℓ^m is $\frac{\ell^{m+1}-1}{\ell-1}$, and $\ell^m + \ell^{m-1}$ of these subgroups are cyclic.*

Proof. Every subgroup of $\mathbb{Z}/\ell^m\mathbb{Z} \times \mathbb{Z}/\ell^m\mathbb{Z}$ is isomorphic to $\mathbb{Z}/\ell^i\mathbb{Z} \times \mathbb{Z}/\ell^j\mathbb{Z}$ for $0 \le i \le j \le m$. The number of subgroups which are isomorphic to $\mathbb{Z}/\ell^i\mathbb{Z} \times \mathbb{Z}/\ell^j\mathbb{Z}$ is 1 if $i = j$ and $\ell^{j-i} + \ell^{j-i-1}$ otherwise.

A direct consequence of the above statement is that there are

$$\sum_{i=0}^{\lfloor \frac{m-1}{2} \rfloor} \ell^{m-2i} + \ell^{m-2i-1} + \epsilon_m = \sum_{t=0}^{m} \ell^t$$

subgroups, where $\epsilon_m = 0$ if k is odd and 1 otherwise. This proves the first statement.

For the second statement, let H be a cyclic subgroup of $\mathbb{Z}/\ell^m\mathbb{Z} \times \mathbb{Z}/\ell^m\mathbb{Z}$ of order l^m. Then H is generated by an element of $\mathbb{Z}/\ell^m\mathbb{Z} \times \mathbb{Z}/\ell^m\mathbb{Z}$ of order l^m, and contains $l^m - l^{m-1}$ elements of order l^m. Therefore, the number of such subgroups is the number of elements of $\mathbb{Z}/\ell^m\mathbb{Z} \times \mathbb{Z}/\ell^m\mathbb{Z}$ of order l^m divided by $l^m - l^{m-1}$.

Let (a, b) be an element of $\mathbb{Z}/\ell^m\mathbb{Z} \times \mathbb{Z}/\ell^m\mathbb{Z}$ of order l^m. Then one of a or b has order l^m. If a has order l^m, then there are $\varphi(\ell^m) = l^m - l^{m-1}$ choices for a, and l^m for b. That is, there are $l^m \cdot (l^m - l^{m-1})$ choices in total.

Otherwise, there are l^{m-1} choices for a (representing the number of elements of order at most l^{m-1}), and $l^m - l^{m-1}$ choices for b. That is, there are $l^{m-1} \cdot (l^m - l^{m-1})$ choices in total. This means the total number of cyclic subgroups of $\mathbb{Z}/\ell^m\mathbb{Z} \times \mathbb{Z}/\ell^m\mathbb{Z}$ of order l^m is

$$\frac{l^m \cdot (l^m - l^{m-1}) + l^{m-1} \cdot (l^m - l^{m-1})}{l^m - l^{m-1}} = l^m + l^{m-1}.$$

□

Remark 4.7. One could also see the first statement in the lemma above by noting that this is the same as the degree of the Hecke operator T_{ℓ^m} which is $\sigma_1(\ell^m)$. We thank the referee for pointing this out.

Corollary 4.8. There are $\frac{\ell^{m+1}-1}{\ell-1}$ separable ℓ^m-isogenies originating at a fixed elliptic curve, and $\ell^m + \ell^{m-1}$ of them are cyclic. (Here we are counting isogenies as different if they differ even after composition with any automorphism of the image.)

Using the correspondence from Corollary 4.5, we then obtain the following.

Theorem 4.9. The number of chains of ℓ-isogenies of length m without backtracking is $\ell^m + \ell^{m-1}$. (Here we do not distinguish between isogenies that differ by composition with an automorphism on the image.)

This last result can be observed in a much more elementary way, which is also enlightening. We consider chains of ℓ-isogenies of length m. To analyze the situation, it is helpful to draw a graph similar to an ℓ-isogeny graph but that does *not* identify isomorphic curves. This graph is an $(\ell + 1)$-regular tree of depth m. The root of the tree has $\ell + 1$ children, and every other node (except the leaves) has ℓ children. The leaves have depth m. It is easy to work out that the number of leaves in this tree is $(\ell + 1)\ell^{m-1}$, and this is also equal to the number of paths of length m without backtracking, as stated in Theorem 4.9.

Finally, this graph also helps us count the number of chains of ℓ-isogenies of length m including those that backtrack. By examining the graph carefully, we can see that the number of such walks is $\ell^m + \ell^{m-1} + \ldots + \ell + 1$, and according to Corollary 4.8, this corresponds to the number of ℓ^m-isogenies that are not necessarily cyclic.

These results were also observed experimentally using Sage. The numbers match the results of our experiments for small values of ℓ and m, over various finite fields and for different choices of elliptic curves, see Table 1. Notice that the images under isogenies with distinct kernels may be isomorphic, leading to double edges in an isogeny graph that identifies isomorphic curves. Hence, the number of isomorphism classes of images (i.e., the number of neighbors in the isogeny graph) may be smaller than the number of isogenies stated in the table.

Table 1 For small fixed ℓ and m, values obtained experimentally for the number of ℓ-isogeny-chains of length m starting at a fixed elliptic curve E without and with backtracking.

ℓ	m	Number of isogenies without backtracking	Number of isogenies with backtracking
2	4	24	31
2	5	48	63
2	6	96	127
2	7	192	255
3	4	108	121
3	5	324	364

Part 2: Constructions of Ramanujan Graphs

In this section we review the constructions of two families of Ramanujan graph, LPS graphs and Pizer graphs. Ramanujan graphs are optimal expanders; see Section 5 for some related background. The purpose is twofold. On the one hand, we wish to explain how equivalent constructions on the same object highlight different significant properties. On the other hand, we wish to explicate the relationship between LPS graphs and Pizer graphs.

Both families (LPS and Pizer) of Ramanujan graphs can be viewed (cf. [12, Section 3]) as a set of "local double cosets," i.e., as a graph on

$$\Gamma \backslash \mathrm{PGL}_2(\mathbb{Q}_l)/\mathrm{PGL}_2(\mathbb{Z}_l), \tag{3}$$

where Γ is a discrete cocompact subgroup. In both cases, one has a chain of isomorphisms that are used to show these graphs are Ramanujan, and in both cases one may in fact vary parameters to get an infinite family of Ramanujan graphs.

To explain this better, we introduce some notation. Let us choose a pair of distinct primes p and l for an $(l + 1)$-regular graph whose size depends on p. (An infinite family of Ramanujan graphs is formed by varying p.) Let us fix a quaternion algebra B defined over $\mathbb{Q}$ and ramified at exactly one finite prime and at ∞, and an order of the quaternion algebra $\mathcal{O}$. Let $\mathbb{A}$ denote the adèles of $\mathbb{Q}$ and $\mathbb{A}_f$ denote the finite adèles. For precise definitions see Section 5.

In the case of Pizer graphs, let $B = B_{p,\infty}$ be ramified at p and ∞, and take $\mathcal{O}$ to be a maximal order (i.e., an order of level p).[1] Then we may construct (as in [20]) a graph by giving its adjacency matrix as a Brandt matrix. (The Brandt matrix is given via an explicit matrix representation of a Hecke operator associated to $\mathcal{O}$.) Then we have (cf. [4, (1)]) a chain of isomorphisms connecting (3) with supersingular isogeny graphs (SSIG) discussed in Part 1 above:

$$(\mathcal{O}[l^{-1}])^{\times} \backslash \mathrm{GL}_2(\mathbb{Q}_l)/\mathrm{GL}_2(\mathbb{Z}_l) \cong B^{\times}(\mathbb{Q}) \backslash B^{\times}(\mathbb{A}_f)/B^{\times}(\hat{\mathbb{Z}}) \cong \mathrm{Cl}\mathcal{O} \cong \mathrm{SSIG}. \tag{4}$$

This can be used (cf. [4, 5.3.1]) to show that the supersingular l-isogeny graph is connected, as well as the fact that it is indeed a Ramanujan graph.

In the case of LPS graphs the choices are very different. Let $B = B_{2,\infty}$ now be the Hamiltonian quaternion algebra. The group Γ in (3) is chosen as a congruence

---

[1] A similar construction exists for a more general $\mathcal{O}$. However, to relate the resulting graph to supersingular isogeny graphs, we require $\mathcal{O}$ to be maximal.

subgroup dependent on p. This leads to a larger graph whose constructions fit into the following chain of isomorphisms:

$$PSL_2(\mathbb{F}_p) \cong \Gamma(2p)\backslash\Gamma(2) \cong \Gamma(2p)\backslash T$$

$$\cong \Gamma(2p)\backslash PGL_2(\mathbb{Q}_l)/PGL_2(\mathbb{Z}_l) \cong G'(\mathbb{Q})\backslash H_{2p}/G'(\mathbb{R})K_0^{2p}. \quad (5)$$

The isomorphic constructions and their relationship will be made explicit in Sections 6.1–6.3 and Section 7.2. We shall also explain how properties of the graph, such as its regularity, connectedness, and the Ramanujan property, are highlighted by this chain of isomorphisms. For now we give only an overview, to be able to compare this case with that of Pizer graphs. The quotient $PGL_2(\mathbb{Q}_l)/PGL_2(\mathbb{Z}_l)$ has a natural structure of an infinite tree T. This tree can be defined in terms of homothety classes of rank two lattices of $\mathbb{Q}_l \times \mathbb{Q}_l$ (see Section 6.2). One may define a group $G' = B^\times/Z(B^\times)$ and its congruence subgroups $\Gamma(2)$ and $\Gamma(2p)$, and show that the discrete group $\Gamma(2)$ acts simply transitively on the tree T, and hence $\Gamma(2p)\backslash T$ is isomorphic to the finite group $\Gamma(2)/\Gamma(2p)$. Using the Strong Approximation theorem, this turns out to be isomorphic to the group $PSL_2(\mathbb{F}_p)$. The latter has a structure of an $(l+1)$-regular Cayley graph. A second application of the Strong Approximation theorem with K_0^{2p}, an open compact subgroup of $G'(\mathbb{A}_f)$, shows that H_{2p} is a finite index normal subgroup of $G'(\mathbb{A})$.

Note that an immediate distinction between Pizer and LPS graphs is that the quaternion algebras underlying the constructions are different: they ramify at different finite primes (p and 2, respectively). In addition, the size of the discrete subgroup Γ determining the double cosets of (3) is different in the two cases. Accordingly, the size of the resulting graphs is different as well. We shall see that (under appropriate assumptions on p and l) the Pizer graph has $\frac{p-1}{12}$ vertices, while the LPS graph has order $|PSL_2(\mathbb{F}_p)| = \frac{p(p^2-1)}{2}$. One may consider an order $\mathcal{O}_{LPS}$ such that $(\mathcal{O}_{LPS}[l^{-1}])^\times \cong \Gamma(2p)$ analogously to the relationship of $\mathcal{O}$ and Γ in the Pizer case and (4). However, this order $\mathcal{O}_{LPS}$ is unlike the Eichler order from the Pizer case. (It has a much higher level.) In particular, there is a discrepancy between the order of the class set $Cl\mathcal{O}_{LPS}$ and the order of the LPS graph. This is a numerical obstruction indicating that an analogue of the chain (4) for LPS graphs is at the very least not straightforward.

The rest of the paper has the following outline. In Section 6 we explore the isomorphic constructions of LPS graphs from (5). We give the construction as a Cayley graph in Section 6.1. The infinite tree of homothety classes of lattices is given in Section 6.2. In Section 6.3 we explain how local double cosets of the Hamiltonian quaternion algebra connect these constructions. Section 6.4 makes one step of the chain of isomorphisms in (5) completely explicit in the case of $l = 5$ and $l = 13$, and describes how the same can be done in general. In Section 7 we give an overview of how Strong Approximation plays a role in proving the isomorphisms and the connectedness and Ramanujan property of the graphs. In Section 8 we turn briefly to Pizer graphs. We summarize the construction, and explain how various restrictions on the prime p guarantee properties of the graph. Section 8.2 contains the computation of a prime p where the existence of both an LPS and a Pizer construction is guaranteed (for $l = 5$). In Section 9 we say a bit more of the relationship of Pizer and LPS graphs, having introduced more of the objects mentioned in passing above.

Throughout this part of the paper we aim to only include technical details if we can make them fairly self-contained and explicit, and otherwise to give a reference for further information.

5   Background on Ramanujan Graphs and Adèles

In this section we fix notation and review some definitions and facts that we will be using for the remainder of Part 2.

Expander graphs are graphs where small sets of vertices have many neighbors. For many applications of expander graphs, such as in Part 1, one wants $(l + 1)$-regular expander graphs X with l small and the number of vertices of X large. If X is an $(l + 1)$-regular graph (i.e., where every vertex has degree $l + 1$), then $l + 1$ is an eigenvalue of the adjacency matrix of X. All eigenvalues λ satisfy $-(l + 1) \leq \lambda \leq (l + 1)$, and $-(l + 1)$ is an eigenvalue if and only if X is bipartite. Let $\lambda(X)$ be the second largest eigenvalue in absolute value of the adjacency matrix. The smaller $\lambda(X)$ is, the better expander X is. Alon–Boppana proved that for an *infinite* family of $(l + 1)$-regular graphs of increasing size, $\liminf_{(X)} \lambda(X) \geq 2\sqrt{l}$ [2]. An $(l + 1)$-regular graph X is called Ramanujan if $\lambda(X) \leq 2\sqrt{l}$. Thus an infinite family of Ramanujan graphs are optimal expanders.

For a finite prime p, let $\mathbb{Q}_p$ denote the field of p-adic numbers and $\mathbb{Z}_p$ its ring of integers. Let $\mathbb{Q}_\infty = \mathbb{R}$. We denote the adèle ring of $\mathbb{Q}$ by $\mathbb{A}$ and recall that it is defined as a restricted direct product in the following way:

$$\mathbb{A} = \prod_p{}' \mathbb{Q}_p = \left\{ (a_p) \in \prod_p \mathbb{Q}_p : a_p \in \mathbb{Z}_p \text{ for all but a finite number of } p < \infty \right\}.$$

We denote the ring of finite adèles by $\mathbb{A}_f$, that is

$$\mathbb{A}_f = \prod_{p<\infty}{}' \mathbb{Q}_p = \left\{ (a_p) \in \prod_{p<\infty} \mathbb{Q}_p : a_p \in \mathbb{Z}_p \text{ for all but a finite number of } p \right\}.$$

Let $\mathbb{A}^\times$ denote the idèle group of $\mathbb{Q}$, the group of units of $\mathbb{A}$,

$$\mathbb{A}^\times = \prod_p{}' \mathbb{Q}_p = \left\{ (a_p) \in \prod_p \mathbb{Q}_p^\times : a_p \in \mathbb{Z}_p^\times \text{ for all but a finite number of } p < \infty \right\}.$$

Let B be a quaternion algebra over $\mathbb{Q}$, $B^\times$ the invertible elements of B and $\mathcal{O}$ an order of B. For a prime p let $\mathcal{O}_p = \mathcal{O} \otimes_\mathbb{Z} \mathbb{Z}_p$. Then let

$$B^\times(\mathbb{A}) = \prod_p{}' B^\times(\mathbb{Q}_p)$$

$$= \left\{ (g_p) \in \prod_p B^\times(\mathbb{Q}_p) : g_p \in \mathcal{O}_p^\times \text{ for all but a finite number of } p < \infty \right\}.$$

More generally for an indexed set of locally compact groups $\{G_v\}_{v \in I}$ with a corresponding indexed set of compact open subgroups $\{K_v\}_{v \in I}$ we may define the restricted direct product of the G_v with respect to the K_v by the following:

$$G := \prod_{v \in I}{}' G_v = \left\{ (g_v) \in \prod_{v \in I} G_v : g_v \in K_v \text{ for all but a finite number of } v \right\}.$$

If we define a neighborhood base of the identity as

$$\left\{ \prod_v U_v : U_v \text{ neighborhood of identity in } G_v \right.$$

$$\left. \text{and } U_v = K_v \text{ for all but a finite number of } v \right\}$$

then G is a locally compact topological group.

6 LPS Graphs

We describe the LPS graphs used in [3] for a proposed hash function. They were first considered in [14], for further details see also [15]. We shall examine the objects and isomorphisms in (5) in more detail. We review constructions of these graphs in turn as Cayley graphs and graphs determined by rank two lattices or, equivalently, local double cosets. Throughout this section, let l and p be distinct, odd primes both congruent to 1 modulo 4. We shall give constructions of $(l + 1)$-regular Ramanujan graphs whose size depends on p. We shall also assume for convenience[2] that $\left(\frac{p}{l}\right) = 1$, i.e., that p is a square modulo l.

---

[2] If p is not a square modulo l, then the constructions described below result in bipartite Ramanujan graphs with twice as many vertices.

6.1 Cayley Graph over $\mathbb{F}_p$.

This description follows [14, Section 2]. The graph we are interested in is the Cayley graph of the group $PSL_2(\mathbb{F}_p)$. We specify a set of generators S below. The vertices of the graph are the $\frac{p(p^2-1)}{2}$ elements of $PSL_2(\mathbb{F}_p)$. Two vertices $g_1, g_2 \in PSL_2(\mathbb{F}_p)$ are connected by an edge if and only if $g_2 = g_1 h$ for some $h \in S$.

Next we give the set of generators S. Since $l \equiv 1 \mod 4$ it follows from a theorem of Jacobi [15, Theorem 2.1.8] that there are $l + 1$ integer solutions to

$$l = x_0^2 + x_1^2 + x_2^2 + x_3^2; \quad 2 \nmid x_0; \quad x_0 > 0. \tag{6}$$

In this case we will also have $2|x_i$ for all $i > 0$. Let S be the set of solutions of (6). Since $p \equiv 1 \mod 4$ we have $\left(\frac{-1}{p}\right) = 1$. Let $\varepsilon \in \mathbb{Z}$ such that $\varepsilon^2 \equiv -1 \mod p$. Then to each solution of (6) we assign an element of $PGL_2(\mathbb{Z})$ as follows:

$$(x_0, x_1, x_2, x_3) \mapsto \begin{pmatrix} x_0 + x_1\varepsilon & x_2 + x_3\varepsilon \\ -x_2 + x_3\varepsilon & x_0 - x_1\varepsilon \end{pmatrix}. \tag{7}$$

Note that the matrix on the right-hand side has determinant $l \mod p$. Since $\left(\frac{l}{p}\right) = 1$ this determines an element of $PSL_2(\mathbb{F}_p)$. The $l + 1$ elements of $PSL_2(\mathbb{F}_p)$ determined by (7) form the set of Cayley generators. Let us abuse notation and denote this set with S as well. This graph is connected. To prove this fact, one may use the theory of quadratic Diophantine equations [14, Proposition 3.3]. Alternately, the chain of isomorphisms (5) proves this fact by relating this Cayley graph to a quotient of a connected graph [15, Theorem 7.4.3]: the infinite tree we shall describe in the next section.

The solutions (x_0, x_1, x_2, x_3) and $(x_0, -x_1, -x_2, -x_3)$ correspond to elements of S that are inverses in $PSL_2(\mathbb{F}_p)$. Since $|S| = l + 1$ this implies that the generators determine an undirected $(l + 1)$-regular graph.

6.2 Infinite Tree of Lattices

Next we shall work over $\mathbb{Q}_l$. We give a description of the same graph in two ways: in terms of homothety classes of rank two lattices, and in terms of local double cosets of the multiplicative group of the Hamiltonian quaternion algebra. The description follows [15, 5.3, 7.4]. Let $B = B_{2,\infty}$ be the Hamiltonian quaternion algebra defined over $\mathbb{Q}$.

First we review the construction of an $(l + 1)$-regular infinite tree on homothety classes of rank two lattices in $\mathbb{Q}_l \times \mathbb{Q}_l$ following [15, 5.3]. The vertices of this infinite graph are in bijection with $PGL_2(\mathbb{Q}_l)/PGL_2(\mathbb{Z}_l)$. To talk about a finite graph, we shall then consider two subgroups $\Gamma(2)$ and $\Gamma(2p)$ in $B^\times/Z(B^\times)$. It turns out that

$\Gamma(2)$ acts simply transitively on the infinite tree, and orbits of $\Gamma(2p)$ on the tree are in bijection with the finite group $\Gamma(2)/\Gamma(2p)$. Under our assumptions the latter turns out to be in bijection with $\mathrm{PSL}_2(\mathbb{F}_p)$ above and the finite quotient of the tree is isomorphic to the Cayley graph above.

First we describe the infinite tree following [15, 5.3]. Consider the two-dimensional vector space $\mathbb{Q}_l \times \mathbb{Q}_l$ with standard basis $\mathbf{e}_1 = {}^t\langle 1, 0\rangle$, $\mathbf{e}_2 = {}^t\langle 0, 1\rangle$. A *lattice* is a rank two $\mathbb{Z}_l$-submodule $L \subset \mathbb{Q}_l \times \mathbb{Q}_l$. It is generated (as a $\mathbb{Z}_l$-module) by two column vectors $\mathbf{u}, \mathbf{v} \in \mathbb{Q}_l \times \mathbb{Q}_l$ that are linearly independent over $\mathbb{Q}_l$. We shall consider homothety classes of lattices, i.e., we say lattices L_1 and L_2 are equivalent if there exists an $0 \neq \alpha \in \mathbb{Q}_l$ such that $\alpha L_1 = L_2$. Writing $\mathbf{u}, \mathbf{v}$ in the standard basis $\mathbf{e}_1, \mathbf{e}_2$ maps the lattice L to an element $M_L \in \mathrm{GL}_2(\mathbb{Q}_l)$. Let $\mathbf{u}_1, \mathbf{v}_1, \mathbf{u}_2, \mathbf{v}_2 \in \mathbb{Q}_l \times \mathbb{Q}_l$ and let $L_i = \mathrm{Span}_{\mathbb{Z}_l}\{\mathbf{u}_i, \mathbf{v}_i\}$ ($i = 1, 2$) be the lattices generated by these respective pairs of vectors, with M_{L_1} and M_{L_2} the corresponding matrices. Let $M \in \mathrm{GL}_2(\mathbb{Q}_l)$ so that $M_{L_1}M = M_{L_2}$. Then $L_1 = L_2$ (as subsets of $\mathbb{Q}_l \times \mathbb{Q}_l$) if and only if $M \in \mathrm{GL}_2(\mathbb{Z}_l)$. It follows that the homothety classes of lattices are in bijection with $\mathrm{PGL}_2(\mathbb{Q}_l)/\mathrm{PGL}_2(\mathbb{Z}_l)$. Equivalently, we may say that $\mathrm{PGL}_2(\mathbb{Q}_l)/\mathrm{PGL}_2(\mathbb{Z}_l)$ acts simply transitively on homothety classes of lattices.

The vertices of the infinite graph T are homothety classes of lattices. The classes $[L_1], [L_2]$ are adjacent in T if and only if there are representatives $L'_i \in [L_i]$ ($i = 1, 2$) such that $L'_2 \subset L'_1$ and $[L'_1 : L'_2] = l$. We show that this relation defines an undirected $(l + 1)$-regular graph. By the transitive action of $\mathrm{GL}_2(\mathbb{Q}_l)$ on lattices we may assume that $L'_1 = \mathbb{Z}_l \times \mathbb{Z}_l = \mathrm{Span}_{\mathbb{Z}_l}\{\mathbf{e}_1, \mathbf{e}_2\}$, the *standard lattice* and $L'_2 \subset \mathbb{Z}_l \times \mathbb{Z}_l$. The map $\mathbb{Z}_l \to \mathbb{Z}_l/l\mathbb{Z}_l \cong \mathbb{F}_l$ induces a map from $\mathbb{Z}_l \times \mathbb{Z}_l$ to $\mathbb{F}_l^2$. Since the index of L'_2 in $\mathbb{Z}_l \times \mathbb{Z}_l$ is l, the image of L'_2 is a one-dimensional vector subspace of $\mathbb{F}_l^2$. This implies that $L'_2 \supset \{l\mathbf{e}_1, l\mathbf{e}_2\}$, i.e., $L'_2 \supset lL'_1$ and the graph is undirected.[3] Furthermore, since there are $l + 1$ one-dimensional subspaces of $\mathbb{F}_l^2$, the graph is $(l + 1)$-regular.

The $l + 1$ neighbors of the standard lattice can be described explicitly by the following matrices:

$$M_l = \begin{pmatrix} 1 & 0 \\ 0 & l \end{pmatrix}, \quad M_h = \begin{pmatrix} l & h \\ 0 & 1 \end{pmatrix} \text{ for } 0 \leq h \leq l - 1 \tag{8}$$

For any of the matrices M_t ($0 \leq t \leq l$) the columns of M_t span a different one-dimensional subspace of $\mathbb{F}_l \times \mathbb{F}_l$. The matrices determine the neighbors of any other lattice by a change of basis in $\mathbb{Q}_l \times \mathbb{Q}_l$.

By the above we can already see that T is isomorphic to the graph on $\mathrm{PGL}_2(\mathbb{Q}_l)/\mathrm{PGL}_2(\mathbb{Z}_l)$ with edges corresponding to multiplication by generators (8) above. To show that T is a tree it suffices to show that there is exactly one path from the standard lattice $\mathbb{Z}_l \times \mathbb{Z}_l$ to any other homothety class. This follows from the uniqueness of the Jordan–Hölder series in a finite cyclic l-group as in [15, p. 69].

---

[3]That is, the adjacency relation defined above is symmetric.

In the next section, we show that the above infinite tree is isomorphic to a Cayley graph of a subgroup of $B^\times/Z(B^\times)$. In Section 6.4 we give an explicit bijection between the Cayley generators and the matrices given in (8) above.

6.3 Hamiltonian Quaternions over a Local Field

To turn the above infinite tree into a finite, $(l + 1)$-regular graph we shall define a group action on its vertices. Let B be the algebra of Hamiltonian quaternions defined over $\mathbb{Q}$. Let G' be the $\mathbb{Q}$-algebraic group $B^\times/Z(B^\times)$. In this subsection we shall follow [15, 7.4] to define normal subgroups $\Gamma(2p) \subset \Gamma(2)$ of $\Gamma = G'(\mathbb{Z}[l^{-1}])$ such that $\Gamma(2)$ acts simply transitively on the graph T. The quotient $\Gamma(2p)\backslash T$ will be isomorphic to the Cayley graph of the finite quotient group $\Gamma(2)/\Gamma(2p)$. This graph is isomorphic to the Cayley graph of $\mathrm{PSL}_2(\mathbb{F}_p)$ defined in Section 6.1 above. Thus we have the following equation.

$$\mathrm{PSL}_2(\mathbb{F}_p) \cong \Gamma(2p)\backslash\Gamma(2) \cong \Gamma(2p)\backslash T \cong \Gamma(2p)\backslash \mathrm{PGL}_2(\mathbb{Q}_l)/\mathrm{PGL}_2(\mathbb{Z}_l). \tag{9}$$

We first define the groups Γ, $\Gamma(2)$, $\Gamma(2p)$ and then examine their relationship with T. Recall that $B = B_{2,\infty}$, i.e., B is ramified at 2 and ∞. For a commutative ring R define $B(R) = \mathrm{Span}_R\{1, \mathbf{i}, \mathbf{j}, \mathbf{k}\}$ where $\mathbf{i}^2 = \mathbf{j}^2 = -1$ and $\mathbf{ij} = -\mathbf{ji} = \mathbf{k}$. We introduce the notation $b_{x_0,x_1,x_2,x_3} := x_0 + x_1\mathbf{i} + x_2\mathbf{j} + x_3\mathbf{k}$. Recall that for $b = b_{x_0,x_1,x_2,x_3}$ we may define $\bar{b} = b_{x_0,-x_1,-x_2,-x_3}$ and the *reduced norm* of b as $N(b) = b\bar{b} = x_0^2 + x_1^2 + x_2^2 + x_3^2$. For a (commutative, unital) ring R an element $b \in B(R)$ is invertible in $B(R)$ if and only if $N(b)$ is invertible in R. (Then $b^{-1} = (N(b))^{-1}\bar{b}$.) Furthermore

$$[b_{x_0,x_1,x_2,x_3}, b_{y_0,y_1,y_2,y_3}] = 2(x_2y_3 - x_3y_2)\mathbf{i} + 2(x_3y_1 - x_1y_3)\mathbf{j} + 2(x_1y_2 - x_2y_1)\mathbf{k}, \tag{10}$$

and hence if R has no zero divisors, then $Z(B(R)) = R$. In particular $Z(B^\times(\mathbb{Z}[l^{-1}])) = \{\pm l^k \mid k \in \mathbb{Z}\}$.

Recall that S was the set of $l + 1$ integer solutions of (6). Any solution x_0, x_1, x_2, x_3 determines a $b = b_{x_0,x_1,x_2,x_3} \in B(\mathbb{Z}[l^{-1}])$ such that $N(b) = l$. Since l is invertible in $\mathbb{Z}[l^{-1}]$ we in fact have $b \in B^\times(\mathbb{Z}[l^{-1}])$. Let $\Gamma = G'(\mathbb{Z}[l^{-1}]) = B^\times(\mathbb{Z}[l^{-1}])/Z(B^\times(\mathbb{Z}[l^{-1}]))$ and let us denote the image of S in Γ by S as well. Since $B^\times(\mathbb{Z}[l^{-1}]) = \{b \in B(\mathbb{Z}[l^{-1}]) \mid N(b) = l^k, k \in \mathbb{Z}\}$, if $[b] \in \Gamma$ for $b \in B^\times(\mathbb{Z}[l^{-1}])$, then it follows from [15, Corollary 2.1.10] that b is a unit multiple of an element of $\langle S\rangle$. It follows that $\Gamma = \langle S\rangle\{[1], [\mathbf{i}], [\mathbf{j}], [\mathbf{k}]\}$ and the index of $\langle S\rangle$ in Γ is 4. In fact observe that if $b \in S$ then $b^{-1} \in S$ and [15, Corollary 2.1.11] states that $\langle S\rangle$ is a free group on $\frac{l+1}{2}$ generators. We shall see that $\langle S\rangle$ agrees with a congruence subgroup $\Gamma(2)$.

Now let $N = 2M$ be coprime to l and let $R = \mathbb{Z}[l^{-1}]/N\mathbb{Z}[l^{-1}]$. The quotient map $\mathbb{Z}[l^{-1}] \to R$ determines a map $B(\mathbb{Z}[l^{-1}]) \to B(R)$. This restricts to a map $B^\times(\mathbb{Z}[l^{-1}]) \to B^\times(R)$. Observe that if $M = 1$ then $B^\times(R)$ is commutative. If

$M = p$, then the subgroup

$$Z := \left\{ b_{x_0,0,0,0} \in B^\times(\mathbb{Z}[l^{-1}]/2p\mathbb{Z}[l^{-1}]) \mid p \nmid x_0, 2 \nmid x_0 \right\}$$

(cf. [14, p. 266]) is central in $B^\times(R)$. Consider the commutative diagram:

$$
\begin{array}{ccccc}
B(\mathbb{Z}[l^{-1}])^\times & \longrightarrow & B^\times(\mathbb{Z}[l^{-1}]/2\mathbb{Z}[l^{-1}]) & \longrightarrow & B^\times(\mathbb{Z}[l^{-1}]/2p\mathbb{Z}[l^{-1}]) \\
\downarrow & & \downarrow & & \downarrow \\
\Gamma & \xrightarrow{\pi_2} & B^\times(\mathbb{Z}[l^{-1}]/2\mathbb{Z}[l^{-1}]) & \xrightarrow{\pi_p} & B^\times(\mathbb{Z}[l^{-1}]/2p\mathbb{Z}[l^{-1}])/Z
\end{array}
\tag{11}
$$

and define[4] $\pi_{2p} := \pi_p \circ \pi_2$ and $\Gamma(2) := \ker \pi_2$ and $\Gamma(2p) = \ker \pi_{2p}$. Observe that by the congruence conditions (cf. (6)) $S \subseteq \Gamma$ is contained in $\Gamma(2)$ and in fact $\langle S \rangle = \Gamma(2) \supseteq \Gamma(2p)$. As mentioned above this implies that $\Gamma(2)$ is a free group with $\frac{l+1}{2}$ generators.

To see the action of $\Gamma(2)$ on T note that B splits over $\mathbb{Q}_l$ and hence $B(\mathbb{Q}_l) \cong M_2(\mathbb{Q}_l)$. Since $-1 \in (\mathbb{F}_l^\times)^2$ there exists an $\epsilon \in \mathbb{Z}_l$ such that $\epsilon^2 = -1$. Then we have an isomorphism $\sigma : B(\mathbb{Q}_l) \to M_2(\mathbb{Q}_l)$ [15, p. 95] given by

$$\sigma(x_0 + x_1\mathbf{i} + x_2\mathbf{j} + x_3\mathbf{k}) = \begin{pmatrix} x_0 + x_1\epsilon & x_2 + x_3\epsilon \\ -x_2 + x_3\epsilon & x_0 - x_1\epsilon \end{pmatrix}. \tag{12}$$

Observe that $\sigma(B^\times(\mathbb{Z}[l^{-1}])) \subseteq GL_2(\mathbb{Q}_l)$ and σ maps elements of the center into scalar matrices, and hence this defines an action of Γ (and hence $\Gamma(2), \Gamma(2p)$) on T. This action preserves the graph structure. Then we have the following. Observe that σ maps the elements of $\langle S \rangle \subseteq \Gamma$ into the congruence subgroup of $PGL_2(\mathbb{Z}_l)$ modulo 2.

Proposition 6.1 ([15, Lemma 7.4.1]). *The action of $\Gamma(2)$ on the tree $T = PGL_2(\mathbb{Q}_l)/PGL_2(\mathbb{Z}_l)$ is simply transitive (and respects the graph structure).*

Proof. See *loc.cit.* for details of the proof. Transitivity follows from the fact that T is connected and elements of S map a vertex of T to its distinct neighbors. The group $\Gamma(2) = \langle S \rangle$ is a discrete free group, hence its intersection with a compact stabilizer $PGL_2(\mathbb{Z}_l)$ is trivial. This implies that the neighbors are distinct and the stabilizer of any vertex is trivial. $\square$

The above implies that the orbits of $\Gamma(2p)$ on T have the structure of the Cayley graph $\Gamma(2)/\Gamma(2p)$ with respect to the generators S. We can see from the maps

---

[4]The definition here agrees with the choices in [14] as well as $\Gamma(N) = \ker(G'(\mathbb{Z}[l^{-1}]) \to G'(\mathbb{Z}[l^{-1}]/N\mathbb{Z}[l^{-1}]))$ in [15]. Here $G' = B^\times/Z(B^\times)$ as a $\mathbb{Q}$-algebraic group. Note however that by (10) the center $Z(B^\times(R))$ for $R = \mathbb{Z}[l^{-1}]/N\mathbb{Z}[l^{-1}]$, $N = 2M$ may not be spanned by $1 + N\mathbb{Z}[l^{-1}]$. In fact from (10) $B^\times(R)$ is commutative for $M = 1$ and for $M = p$ we have $Z(B^\times(R)) = Z \oplus [p]\mathbf{i} + [p]\mathbf{j} + [p]\mathbf{k}$. However the image of $\langle S \rangle$ in $B^\times(R)$ is trivial if $M = 1$ and intersects the center in Z when $M = p$.

in (11) that $\Gamma(2)/\Gamma(2p)$ is isomorphic to a subgroup of $G'(\mathbb{Z}/2p\mathbb{Z}) \cong G'(\mathbb{Z}/2\mathbb{Z}) \times G'(\mathbb{Z}/p\mathbb{Z})$. (This last isomorphism follows from the Chinese Remainder Theorem.) Since the image of $\Gamma(2)$ in $G'(\mathbb{Z}/2\mathbb{Z})$ is trivial, we may identify $\Gamma(2)/\Gamma(2p)$ with a subgroup of $G'(\mathbb{Z}/p\mathbb{Z})$. Here $G'(\mathbb{Z}/p\mathbb{Z}) \cong \mathrm{PGL}_2(\mathbb{F}_p)$. (For an explicit isomorphism take an analogue of σ in (12) with $\epsilon \in \mathbb{Z}/p\mathbb{Z}$ such that $\epsilon^2 = -1$.) The image of $\Gamma(2)$ agrees with $\mathrm{PSL}_2(\mathbb{F}_p)$ as a consequence of the Strong Approximation theorem [15, Lemma 7.4.2]. We shall discuss this in the next section.

We summarize the contents of this section.

Theorem 6.2 ([15, Theorem 7.4.3]). *Let l and p be primes so that $l \equiv p \equiv 1$ mod 4 and l is a quadratic residue modulo $2p$. Let $S \subset \mathrm{PSL}_2(\mathbb{F}_p)$ be the $(l+1)$-element set corresponding to the solutions of (6) via the map (7) and $Cay(\mathrm{PSL}_2(\mathbb{F}_p), S)$ the Cayley graph determined by the set of generators S on the group $\mathrm{PSL}_2(\mathbb{F}_p)$. Let T be the graph on $\mathrm{PGL}_2(\mathbb{Q}_l)/\mathrm{PGL}_2(\mathbb{Z}_l)$ with edges corresponding to multiplication by elements listed in (8). Let B be the Hamiltonian quaternion algebra over $\mathbb{Q}$ and $\Gamma(2p)$ the kernel of the map π_{2p} in (11) (a cocompact congruence subgroup). Then $\Gamma(2p)$ acts on the infinite tree T and we have the following isomorphism of graphs:*

$$Cay(\mathrm{PSL}_2(\mathbb{F}_p), S) \cong \Gamma(2p)\backslash\mathrm{PGL}_2(\mathbb{Q}_l)/\mathrm{PGL}_2(\mathbb{Z}_l). \qquad (13)$$

These are connected, $(l+1)$-regular, non-bipartite, simple, graphs on $\frac{p^3-p}{2}$ vertices.

6.4 Explicit Isomorphism Between Generating Sets

We have seen above that the LPS graph can be interpreted as a finite quotient of the infinite tree of homothety classes of lattices. In this case, the edges are given by matrices that take a $\mathbb{Z}_l$-basis of one lattice to a $\mathbb{Z}_l$-basis of one of its neighbors. On the other hand, the edges can be given in terms of the set of generators S. Proposition 6.1 states that $\langle\sigma(S)\rangle = \Gamma(2) \subset G'(\mathbb{Z}[l^{-1}])$ acts simply transitively on the tree T. The proof of the proposition (cf. [15, Lemma 7.4.1]) implicitly shows that there exists a bijection between elements of $\sigma(S) \subset \mathrm{PGL}_2(\mathbb{Z}_l)$ and the matrices given in (8).

In this section we wish to make this bijection more explicit. For a fixed $\alpha \in S$ we find the matrix from the list (8) determining the same edge of T. As in Section 6.3 we write $\sigma(\alpha) \in \mathrm{PGL}_2(\mathbb{Z}_l)$ for the elements of $\sigma(S)$. This amounts to finding the matrix M from the list in (8) such that $\sigma(\alpha)^{-1}M \in \mathrm{PGL}_2(\mathbb{Z}_l)$.

To pair up matrices from (8) with the corresponding elements of S, we introduce the following notation. Let us number the solutions to $\alpha\bar{\alpha} = l$ as $\alpha_0, \ldots, \alpha_{l-1}, \alpha_l$ so that we have the correspondence $\sigma(\alpha_h)^{-1}M_h \in \mathrm{PGL}_2(\mathbb{Z}_l)$ for $0 \leq h \leq l$. By giving an explicit correspondence, we mean that given an $\alpha \in \sigma^{-1}(S)$, we determine $0 \leq h \leq l$ such that $\alpha = \alpha_h$.

Elements of $\sigma(S) \subset \mathrm{PGL}_2(\mathbb{Z}_l)$ are given in terms of an $\epsilon \in \mathbb{Z}_l$ such that $\epsilon^2 = -1$. Let a, b be the positive integers such that $a^2 + b^2 = l$ and a is odd. Let $0 \le e \le l - 1$ so that $eb = a$. Then in $\mathbb{Z}_l$ we have either $\epsilon \in e + l\mathbb{Z}_l$ and $\epsilon^{-1} = -\epsilon \in -e + l\mathbb{Z}_l$ or $\epsilon \in -e + l\mathbb{Z}_l$ and $\epsilon^{-1} = -\epsilon \in e + l\mathbb{Z}_l$.

Let $\alpha = x_0 + x_1\mathbf{i} + x_2\mathbf{j} + x_3\mathbf{k}$ so that $\sigma(\alpha) \in S$, and a, b, e, ϵ are as above. Let

$$\alpha_h = x_0^{(h)} + x_1^{(h)}\mathbf{i} + x_2^{(h)}\mathbf{j} + x_3^{(h)}\mathbf{k}$$

for $0 \le h \le l$. Here x_0, x_1, x_2, x_3 are integers; it is convenient to think about them (as well as $x_0^{(h)}, x_1^{(h)}, x_2^{(h)}, x_3^{(h)}$ for $0 \le h \le l$) as being in $\mathbb{Z} \subset \mathbb{Z}_l$. Then

$$\sigma(\alpha)^{-1} = \frac{1}{l}\begin{pmatrix} x_0 - x_1\epsilon & -x_2 - x_3\epsilon \\ x_2 - x_3\epsilon & x_0 + x_1\epsilon \end{pmatrix} \tag{14}$$

and

$$\sigma(\alpha)^{-1} \cdot \begin{pmatrix} l & h \\ 0 & 1 \end{pmatrix} = \begin{pmatrix} x_0 - x_1\epsilon & l^{-1}\,(h(x_0 - x_1\epsilon) + (-x_2 - x_3\epsilon)) \\ x_2 - x_3\epsilon & l^{-1}\,(h(x_2 - x_3\epsilon) + (x_0 + x_1\epsilon)) \end{pmatrix}$$

$$\sigma(\alpha)^{-1} \cdot \begin{pmatrix} 1 & 0 \\ 0 & l \end{pmatrix} = \begin{pmatrix} l^{-1}(x_0 - x_1\epsilon) & -x_2 - x_3\epsilon \\ l^{-1}(x_2 - x_3\epsilon) & x_0 + x_1\epsilon \end{pmatrix} \tag{15}$$

Then by (15) we have that $x_0^{(l)} - x_1^{(l)}\epsilon$ and $x_2^{(l)} - x_3^{(l)}\epsilon$ are in $l\mathbb{Z}_l$. Hence $x_0^{(l)} \in x_1^{(l)}\epsilon + l\mathbb{Z}_l$, and thus $(x_0^{(l)})^2 \in (x_1^{(l)}\epsilon)^2 + l\mathbb{Z}_l = -x_1^2 + l\mathbb{Z}_l$, whence $(x_0^{(l)})^2 + (x_1^{(l)})^2 \in l\mathbb{Z}_l$. Note that since $(x_0^{(l)})^2 + (x_1^{(l)})^2 + (x_2^{(l)})^2 + (x_3^{(l)})^2 = l$ and x_0 is positive, this implies that $(x_0^{(l)})^2 + (x_1^{(l)})^2 = l$ and $(x_2^{(l)})^2 + (x_3^{(l)})^2 = 0$, i.e., $x_2^{(l)} = x_3^{(l)} = 0$ and $x_0^{(l)} = a$, $|x_1^{(l)}| = b$. Note that by the assumptions in Section 6.1, $a \pm b\mathbf{i}, a \pm b\mathbf{j}, a \pm b\mathbf{k} \in S$. A straightforward computation now shows the following.

$$\epsilon \in e + l\mathbb{Z}_l \Rightarrow \alpha_l = a + b\mathbf{i},\ \alpha_0 = a - b\mathbf{i},\ \alpha_e = a - b\mathbf{j},$$

$$\alpha_{l-e} = a + b\mathbf{j},\ \alpha_1 = a - b\mathbf{k},\ \alpha_{l-1} = a + b\mathbf{k}$$

$$\epsilon \in -e + l\mathbb{Z}_l \Rightarrow \alpha_l = a - b\mathbf{i},\ \alpha_0 = a + b\mathbf{i},\ \alpha_e = a - b\mathbf{j},$$

$$\alpha_{l-e} = a + b\mathbf{j},\ \alpha_1 = a + b\mathbf{k},\ \alpha_{l-1} = a - b\mathbf{k} \tag{16}$$

Now let us assume that for $\alpha = x_0 + x_1\mathbf{i} + x_2\mathbf{j} + x_3\mathbf{k}$ we have that $x_0 - x_1\epsilon \notin l\mathbb{Z}_l$. This implies that it remains to determine the h such that $\alpha = \alpha_h$ when α is not one of the solutions covered by (6.4). In that case, we may assume $h \notin \{0, 1, e, l-e, l-1, l\}$ and we have

$$h(x_0 - x_1\epsilon) + (-x_2 - x_3\epsilon) \in l\mathbb{Z}_l; \tag{17}$$

$$h(x_2 - x_3\epsilon) + (x_0 + x_1\epsilon) \in l\mathbb{Z}_l. \tag{18}$$

A straightforward computation based on $\alpha\overline{\alpha} = l$ shows that (17) and (18) are satisfied by the same element in $\mathbb{F}_l = \mathbb{Z}/l\mathbb{Z}$. The element

$$\overline{h} = \frac{x_2 + x_3\epsilon}{x_0 - x_1\epsilon} \in \mathbb{F}_l \tag{19}$$

is well defined, since $x_0 - x_1\epsilon \notin l\mathbb{Z}_l$, furthermore, it uniquely determines an $0 \leq h \leq l$. For a fixed α not covered by (6.4), one may thus find h such that $\alpha = \alpha_h$.

We give two explicit examples.

Example 6.3. When $l = 5$, then $a = 1$, $b = 2$ and $e = 3$. Then (20) gives the bijection between the list in (8) and solutions of $\alpha\overline{\alpha} = 5$ in $B(\mathbb{Q}_5)$. In this case the list in (6.4) is exhaustive.

h		0	1	2	3	4	5
$\epsilon \in 3 + 5\mathbb{Z}_5$	α_h	$1 - 2\mathbf{i}$	$1 - 2\mathbf{k}$	$1 + 2\mathbf{j}$	$1 - 2\mathbf{j}$	$1 + 2\mathbf{k}$	$1 + 2\mathbf{i}$
$\epsilon \in 2 + 5\mathbb{Z}_5$		$1 + 2\mathbf{i}$	$1 + 2\mathbf{k}$	$1 + 2\mathbf{j}$	$1 - 2\mathbf{j}$	$1 - 2\mathbf{k}$	$1 - 2\mathbf{i}$

$$\tag{20}$$

Example 6.4. When $l = 13$, we have $a = 3$, $b = 2$, and $e = 8$. The cases listed in (6.4) are no longer exhaustive. The correspondence is given in Table 2.

Table 2 The correspondence when $\epsilon \in 8 + 13\mathbb{Z}_{13}$ (left) and when $\epsilon \in 5 + 13\mathbb{Z}_{13}$ (right).

h	α_h	h	α_h
0	$3 - 2\mathbf{i}$	0	$3 + 2\mathbf{i}$
1	$3 - 2\mathbf{k}$	1	$3 + 2\mathbf{k}$
2	$1 - 2\mathbf{i} - 2\mathbf{j} - 2\mathbf{k}$	2	$1 + 2\mathbf{i} - 2\mathbf{j} + 2\mathbf{k}$
3	$1 - 2\mathbf{i} + 2\mathbf{j} - 2\mathbf{k}$	3	$1 + 2\mathbf{i} + 2\mathbf{j} + 2\mathbf{k}$
4	$1 + 2\mathbf{i} + 2\mathbf{j} + 2\mathbf{k}$	4	$1 - 2\mathbf{i} + 2\mathbf{j} - 2\mathbf{k}$
5	$3 + 2\mathbf{j}$	5	$3 + 2\mathbf{j}$
6	$1 + 2\mathbf{i} - 2\mathbf{j} + 2\mathbf{k}$	6	$1 - 2\mathbf{i} - 2\mathbf{j} - 2\mathbf{k}$
7	$1 + 2\mathbf{i} + 2\mathbf{j} - 2\mathbf{k}$	7	$1 - 2\mathbf{i} + 2\mathbf{j} + 2\mathbf{k}$
8	$3 - 2\mathbf{j}$	8	$3 - 2\mathbf{j}$
9	$1 + 2\mathbf{i} - 2\mathbf{j} - 2\mathbf{k}$	9	$1 - 2\mathbf{i} - 2\mathbf{j} + 2\mathbf{k}$
10	$1 - 2\mathbf{i} - 2\mathbf{j} + 2\mathbf{k}$	10	$1 + 2\mathbf{i} - 2\mathbf{j} - 2\mathbf{k}$
11	$1 - 2\mathbf{i} + 2\mathbf{j} + 2\mathbf{k}$	11	$1 + 2\mathbf{i} + 2\mathbf{j} - 2\mathbf{k}$
12	$3 + 2\mathbf{k}$	12	$3 - 2\mathbf{k}$
13	$3 + 2\mathbf{i}$	13	$3 - 2\mathbf{i}$

7 Strong Approximation

In this section we briefly explain the significance of Strong Approximation to Ramanujan graphs and particularly the LPS graphs above. As discussed in Section 5 we may consider $G(\mathbb{A})$, the adelic points of a linear algebraic group G defined over $\mathbb{Q}$. The group $G(\mathbb{Q})$ embeds diagonally into $G(\mathbb{A})$, and it is a discrete subgroup. The groups $G(\mathbb{Q}_v)$ are also subgroups of $G(\mathbb{A})$, and $G(\mathbb{A})$ has a well-defined projection onto $G(\mathbb{Q}_v)$. Similarly, for a finite set of places S we may take G_S, the direct product of $G(\mathbb{Q}_v)$ for $v \in S$.

Strong Approximation (when it holds) is the statement that for a group G and a finite set of places S the subgroup $G(\mathbb{Q})G_S$ is *dense* in $G(\mathbb{A})$. This implies that

$$G(\mathbb{A}) = G(\mathbb{Q})G_S K \text{ for any open subgroup } K \leq G(\mathbb{A}). \tag{21}$$

For example, Strong Approximation holds for $G = \mathrm{SL}_2$ and any set of places $S = \{v\}$. However, in the form written above it *does not hold* for GL_2 or PGL_2. However one can prove results similar to (21) for GL_2 adding restrictions on the subgroup K:

$$G(\mathbb{A}) = G(\mathbb{Q})G_S K \text{ for an open subgroup } K \leq G(\mathbb{A}) \text{ if } K \text{ is "sufficiently large."} \tag{22}$$

Here we shall have

$$K = \prod_{v \notin S} K_v; \quad K_v \leq G(\mathbb{Z}_v) \tag{23}$$

and the condition of being "sufficiently large" can be made precise by requiring that the determinant map $\det : K_v \to \mathbb{Z}_v^\times$ be surjective for all $v \notin S$.

Strong Approximation holds for the algebraic group of elements of a quaternion algebra of unit norm [26, Théorème 4.3]. We shall use this statement to prove a statement like (22) for the algebraic group of invertible quaternions. A similar statement then holds for $G' = B^\times/Z(B^\times)$ and a subgroup K' that is not quite "large enough." The implications for Pizer graphs and LPS graphs will be discussed in Sections 7.2 and 7.3 below.

These statements coming from Strong Approximation are crucial for proving that the various constructions produce Ramanujan graphs. As seen in Section 5 the Ramanujan property of a graph can be expressed in terms of its eigenvalues. Given a graph (constructed, e.g., via local double cosets as seen above) the Strong Approximation theorem can be used to relate its spectrum to the representation theory of $G(\mathbb{A})$. In that context a theorem of Deligne resolves the issue by proving a special case of the Ramanujan conjecture (see [15, Theorem 6.1.2, Theorem A.1.2, Theorem A.2.14] and [6]).

7.1 Approximation for Invertible Quaternions

The argument below is adapted from [9, Section 3] and [15, 6.3].[5]

Let B be a (definite) quaternion algebra over $\mathbb{Q}$, $B^\times$ its invertible elements and $B^1 = \{b \in B \mid N(b) = 1\}$ its elements of reduced norm 1, recall $N(b) = b\bar{b}$. Let l be a prime where B is split. Then by [26, Théorème 4.3] we have that $B^1(\mathbb{Q})B^1(\mathbb{Q}_l)$ is dense in $B^1(\mathbb{A})$ thus $B^1(\mathbb{A}) = B^1(\mathbb{Q})B^1(\mathbb{Q}_l)K$ for any open subgroup $K \le B^1(\mathbb{A})$. An open subgroup $K \le B^1(\mathbb{A})$ is of the form $K = \prod_v K_v$ where $K_v \le B^1_v$ is open and $K_v = B^1(\mathbb{Z}_v)$ for all but finitely many places v. It follows that given *any* open subgroups $K_v^{(B^1)} \le B^1(\mathbb{Z}_v)$ $(v \ne l)$ such that $K_v^{(B^1)} = B^1(\mathbb{Z}_v)$ for all but finitely many places v we have that

$$B^1(\mathbb{A}) = B^1(\mathbb{Q})B^1(\mathbb{Q}_l) \prod_{v \ne l} K_v^{(B^1)}. \tag{24}$$

To make a similar statement for $B^\times$ it will be necessary to impose a restriction on the open subgroups K_v.

Theorem 7.1. *Let $K_v \le B^\times(\mathbb{Z}_v)$ for every place $l \ne v < \infty$ so that $K_v = B^\times(\mathbb{Z}_v)$ for all but finitely many v, and the norm map $N : K_v \to \mathbb{Z}_v^\times$ is surjective for every place v. Then*

$$B^\times(\mathbb{A}) = B^\times(\mathbb{Q})B^\times(\mathbb{R})B^\times(\mathbb{Q}_l) \prod_{l \ne v < \infty} K_v. \tag{25}$$

Note that by [27, Lemma 13.4.6] the norm map $N : B^\times(\mathbb{Z}_v) \to \mathbb{Z}_v^\times$ is surjective for every nonarchimedean v.

Proof. Let $b \in B^\times(\mathbb{A})$, we need to show b is contained on the right-hand side. To write b as a product according to the right-hand side of (25) we shall use (24), strong approximation for B^1. Observe first that it suffices to show that any $b \in B^\times(\mathbb{A})$ can be written as

$$b = rhk, \text{ where } r \in B^\times(\mathbb{Q}), \ h \in B^1(\mathbb{A}), \text{ and } k \in B^\times(\mathbb{R})B^\times(\mathbb{Q}_l) \prod_{l \ne v < \infty} K_v. \tag{26}$$

This is because the intersections $K_v \cap B^1(\mathbb{Q}_v)$ are open subgroups of $B^1(\mathbb{Z}_v)$ (and $B^\times(\mathbb{Z}_v) \cap B^1(\mathbb{Z}_v) = B^1(\mathbb{Z}_v)$ at all but finitely many places). It thus follows from (24) (choosing $K_v^{(B^1)} := K_v \cap B^1(\mathbb{Q}_v)$) that the factor $h \in B^1(\mathbb{A}) \subseteq B^\times(\mathbb{A})$ from (26) is contained on the right-hand side of (25). It follows that then $b = rhk$ is

---

[5]In fact, since at every split place v we have $B^\times(\mathbb{Q}_v) \cong \mathrm{GL}_2(\mathbb{Q}_v)$ with the reduced norm on $B^\times$ corresponding to the determinant on GL_2 [26, p. 3] this is the "same argument at all but finitely many places."

contained on the right-hand side of (25) as well. (Note that here the factors of h and k belonging to different components $B^\times(\mathbb{Q}_v)$ commute.)

So we must show that any $b \in B^\times(\mathbb{A})$ decomposes as in (26). Let $b = (b_v)_v$ for $b_v \in B^\times(\mathbb{Q}_v)$ and set $n_v := N(b_v)$. For all but finitely many places v we have $b_v \in B^\times(\mathbb{Z}_v)$ and hence $n_v \in \mathbb{Z}_v^\times$. At a finite set T of finite places we may write $n_v \in v^{m_v}\mathbb{Z}_v^\times$. Let us take

$$n_\mathbb{Q} = \prod_{v \in T} v^{m_v}. \tag{27}$$

Then $n_\mathbb{Q} \in \mathbb{Q}_{>0}$, $n_\mathbb{Q} \in \mathbb{Z}_v^\times$ for every $v \notin T$, $v < \infty$ and hence $n_\mathbb{Q}^{-1} n_v \in \mathbb{Z}_v^\times$ for *every* finite place v.

It is a fact that there is an $r \in B^\times(\mathbb{Q})$ such that $N(r) = n_\mathbb{Q}$. Then for this r we have that the norm of $r^{-1}b \in B^\times(\mathbb{A})$ is in $\mathbb{Z}_v^\times$ for every finite place v.

Let us write $(r^{-1}b)_v$ for the component of $r^{-1}b \in B^\times(\mathbb{A})$ at a place v. There exists a $k \in B^\times(\mathbb{R})B^\times(\mathbb{Q}_l)\prod_{l \neq v < \infty} K_v$, $k = (k_v)_v$ such that $k_l = (r^{-1}b)_l$ and $k_\infty = (r^{-1}b)_\infty$ and $N(k_v) = N((r^{-1}b)_v)$ every other place. This follows from the fact that the norm map $N : K_v \to \mathbb{Z}_v^\times$ is surjective.

Now let $h = r^{-1}bk^{-1}$. We show $h \in B^1(\mathbb{A})$. Write $h = (h_v)_v$ for $h_v \in B^\times(\mathbb{Q}_v)$. It follows from the choice of k that h_l and h_∞ are the identity element of $B^\times(\mathbb{Q}_l)$ and $B^\times(\mathbb{R})$, respectively, and $N(h_v) = 1$ at every other place v. This implies that indeed $h \in B^1(\mathbb{A})$. This completes the proof that a decomposition as in (26) exists, and in turn the proof of (25). $\qquad\square$

7.2 Strong Approximation for LPS Graphs

This section is based on [15, 6.3]. (In particular, we recall and elaborate on the proof of the first statements in [15, Proposition 6.3.3] in the special case when $N = 2p$. This is relevant to understanding the last step in (5).) We apply a similar formula to (25) with a particular choice of open subgroups K'_v to prove a statement that relates double cosets such as in (9) to adelic double cosets. Let $B = B_{2,\infty}$ be the algebra of Hamiltonian quaternions, ramified at 2 and ∞. Recall from Section 6.3 that G' is the $\mathbb{Q}$-algebraic group $B^\times/Z(B^\times)$. Let us fix the prime $l \equiv 1 \mod 4$ as in Section 6. In a similar manner to the proof of (25) it follows that

$$G'(\mathbb{A}) = G'(\mathbb{Q})G'(\mathbb{R})G'(\mathbb{Q}_l) \prod_{l \neq v < \infty} G'(\mathbb{Z}_v). \tag{28}$$

Recall that since B splits at l we have $G'(\mathbb{Q}_l) \cong \mathrm{PGL}_2(\mathbb{Q}_l)$. We wish to have a statement similar to (28) above, replacing $G'(\mathbb{Z}_v)$ at $v = 2$ and $v = p$ by congruence subgroups K'_2 and K'_p. (This p is the one fixed above in Section 6.)

Then isomorphism will no longer hold, but the right-hand side will be a finite index normal subgroup of $G'(\mathbb{A})$.

The choice of the smaller subgroups K'_2 and K'_p is as follows. For $v \in \{2, p\}$ let

$$K'_v = \ker\left(G'(\mathbb{Z}_v) \to G'(\mathbb{Z}_v/v\mathbb{Z}_v)\right). \tag{29}$$

Here $\mathbb{Z}_v/v\mathbb{Z}_v = \mathbb{F}_v$ is a finite field, hence $G'(\mathbb{Z}_v/v\mathbb{Z}_v)$ is finite. It follows that the index $[K_v : K'_v]$ is finite. In fact since $B_{2,\infty}$ splits over p we have that $G'(\mathbb{Z}_p/v\mathbb{Z}_p) \cong \mathrm{PGL}_2(\mathbb{F}_p)$, hence $[K_p : K'_p] = p(p^2 - 1)$. At $v = 2$ we have $G'(\mathbb{F}_2) = B^\times(\mathbb{F}_2)$ hence $[K_2 : K'_2] = 8$.

Let us set K'_v as above if $v \in \{2, p\}$ and $K'_v = K_v = G'(\mathbb{Z}_v)$ otherwise, and let us define

$$H_{2p} := \left(G'(\mathbb{Q})G'(\mathbb{R})G'(\mathbb{Q}_l) \prod_{l \neq v < \infty} K'_v\right). \tag{30}$$

By [15, Proposition 6.3.3] Strong Approximation proves that H_{2p} is a finite index normal subgroup of $G'(\mathbb{A})$.

From the definition of H_{2p} in equation (30) we have a surjection from

$$G'(\mathbb{Q}_l) \to G'(\mathbb{Q})\backslash H_{2p}/G'(\mathbb{R}) \prod_{l \neq v < \infty} K'_v.$$

If g_l and $g'_l \in G'(\mathbb{Q}_l)$ are mapped to the same coset on the right-hand side, then there exists $g_q \in G'(\mathbb{Q}), g_r \in G'(\mathbb{R})$ and $k = \prod_{l \neq v < \infty} k_v \in \prod_{l \neq v < \infty} K'_v$ such that $g_l = g_q g'_l g_r k$. This is equivalent to saying $g_l = g_q g'_l$ and $g_q \in K'_v$ for all $l \neq v < \infty$. By the definitions of the K'_vs this last condition implies $g_q \in \Gamma(2p)$. Thus we see that

$$\Gamma(2p)\backslash G'(\mathbb{Q}_l)/G'(\mathbb{Z}_l) \cong G'(\mathbb{Q})\backslash H_{2p}/G'(\mathbb{R}) \prod_{v < \infty} K'_v. \tag{31}$$

Strong approximation in the manner discussed above is used to prove that LPS graphs are Ramanujan. First one shows that the finite $(l+1)$-regular graph $\Gamma(2p)\backslash T$ is Ramanujan if and only if all irreducible infinite-dimensional unramified unitary representations of $\mathrm{PGL}_2(\mathbb{Q}_l)$ that appear in $L^2(PGL_2(\mathbb{Q}_l)/\Gamma(2p))$ are tempered [15, Corollary 5.5.3]. Then by the isomorphism above which follows from Strong Approximation, one can extend a representation ρ'_l of $\mathrm{PGL}_2(\mathbb{Q}_l)$ to an automorphic representation ρ' of $G'(\mathbb{A})$ in $L^2(G'(\mathbb{Q})\backslash G'(\mathbb{A}))$. By the Jacquet–Langlands correspondence, ρ' corresponds to a cuspidal representation ρ of $\mathrm{PGL}_2(\mathbb{A})$ in $L^2(\mathrm{PGL}_2(\mathbb{Q})\backslash \mathrm{PGL}_2(\mathbb{A}))$ such that ρ_v is discrete series for all v where B ramifies (so in our case, 2 and ∞) [15, Theorem 6.2.1]. Finally, Deligne has proved the Ramanujan–Peterson conjecture in this case of holomorphic modular forms [15, Theorem 6.1.2], [6, 7] which says that for ρ a cuspidal representation of

$PGL_2(\mathbb{A})$ in $L^2(PGL_2(\mathbb{Q})\backslash PGL_2(\mathbb{A}))$ with ρ_∞ discrete series, ρ_l is tempered [15, Theorems 7.1.1 and 7.3.1]. Under the Jacquet–Langlands correspondence, the adjacency matrix of our graph X corresponds to the Hecke operator T_l [15, 5.3] and the Ramanujan conjecture is equivalent to saying that $|\lambda| \leq 2\sqrt{l}$ for all of its eigenvalues $\lambda \neq \pm(l+1)$.

7.3 Strong Approximation for Pizer Graphs

Now we turn to discussing how strong approximation is useful in establishing the bijections in (4). In Section 8 we will discuss Pizer's construction of Ramanujan graphs. These graphs are isomorphic to supersingular isogeny graphs. Their vertex set is the class group of a maximal order $\mathcal{O}$ in the quaternion algebra $B_{p,\infty}$. This set is in bijection with an adelic double coset space, which in turn is in bijection with a set of local double cosets.

Let $B = B_{p,\infty}$ be a quaternion algebra (over $\mathbb{Q}$) ramified exactly at ∞ and at a finite prime p. At every finite prime v, $B(\mathbb{Q}_v)$ has a unique maximal order up to conjugation [26, Lemme 1.4]. Given a maximal order $\mathcal{O}$ of B, one may define the adelic group $B^\times(\mathbb{A}_f)$ as a restricted direct product of the groups $B^\times(\mathbb{Q}_v)$ over the finite places, with respect to $\mathcal{O}_v^\times$. (Recall that this means that any element of $B^\times(\mathbb{A}_f)$ is a vector indexed by the finite places v; the component at v is in $B^\times(\mathbb{Q}_v)$ and in fact in $\mathcal{O}_v^\times$ at all but finitely many places.) This adelic object does not in fact depend on the choice of the maximal ideal $\mathcal{O}$. In particular, at any prime $l \neq p$ where B splits we have $B^\times(\mathbb{Q}_l) \cong GL_2(\mathbb{Q}_l)$ and $\mathcal{O}_l^\times \cong GL_2(\mathbb{Z}_l)$.

Let us now fix a prime l where B splits. The same argument as in Section 7.1 works restricted to $B^\times(\mathbb{A}_f)$ (the finite adèles). It follows that we have

$$B^\times(\mathbb{A}_f) = B^\times(\mathbb{Q})B^\times(\mathbb{Q}_l) \prod_{l \neq v < \infty} B^\times(\mathbb{Z}_v). \tag{32}$$

Proposition 7.2. *We have the bijections (cf. [4, (1)])*

$$B^\times(\mathbb{Q})\backslash B^\times(\mathbb{A}_f)/ \prod_{v < \infty} B^\times(\mathbb{Z}_v) \cong (\mathcal{O}(\mathbb{Z}[l^{-1}]))^\times \backslash B^\times(\mathbb{Q}_l)/B^\times(\mathbb{Z}_l)$$
$$\tag{33}$$
$$\cong (\mathcal{O}(\mathbb{Z}[l^{-1}]))^\times \backslash GL_2(\mathbb{Q}_l)/GL_2(\mathbb{Z}_l).$$

Proof. The first bijection follows from (32) and an argument similar to the proof of (31). Indeed, (32) implies that there is a surjection

$$B^\times(\mathbb{Q}_l) \to B^\times(\mathbb{Q})\backslash B^\times(\mathbb{A}_f)/ \prod_{l \neq v < \infty} B^\times(\mathbb{Z}_v). \tag{34}$$

Now two elements $g_l, g_l' \in B^\times(\mathbb{Q}_l)$ land in the same double coset via this bijection if and only if $g_l = g_q g_l' k$ in $B^\times(\mathbb{A}_f)$. Then $g_l = g_q g_l'$ (from equality at the place l) and $g_q \in B^\times(\mathbb{Z}_v)$ (from equality at the places $l \neq v < \infty$). Consider the element $g_q \in B(\mathbb{Q})$, for example, in terms of its coordinates in the standard basis $\{1, \mathbf{i}, \mathbf{j}, \mathbf{k}\}$ of B. Since $g_q \in B^\times(\mathbb{Z}_v)$ we have that $g_q \in \mathcal{O}(\mathbb{Z}[l^{-1}])$, and $g_q \in B^\times(\mathbb{Q}_l)$ implies that in fact $g_q \in (\mathcal{O}(\mathbb{Z}[l^{-1}]))^\times$. This completes the proof of the first bijection in (33).

Now the second bijection follows from the fact that B splits at the prime l and hence $B^\times(\mathbb{Q}_l) \cong GL_2(\mathbb{Q}_l)$ with the unique maximal order $GL_2(\mathbb{Z}_l)$. $\qquad\square$

Finally, we wish to also address the bijection between the adelic double coset object and the class group of the maximal order $\mathcal{O}$. This fact follows from the fact that ideals of $\mathcal{O}$ are locally principal. We omit defining ideals of an order $\mathcal{O}$ or defining the class group here and instead refer the reader to [26, §4], [5, §2.3], or [27]. For the statement about the bijection between the class group $Cl(\mathcal{O})$ and the adelic double cosets in (33) above, see, for example, [5, Theorem 2.6].

8 Pizer Graphs

In this section we give an overview of Pizer's [20] construction of a Ramanujan graph. The graphs constructed by Pizer are isomorphic to the graphs of supersingular elliptic curves over $\mathbb{F}_{p^2}$ [4, Section 2]. These graphs were considered by Mestre [16] and Ihara [10] before (cf. [11]), but Pizer's construction reveals their connection to quaternion algebras, proving their Ramanujan property. In Section 9 we shall compare the resulting graphs to the LPS construction described above.

Pizer's description is in terms of a quaternion algebra and a pair of prime parameters p, l. We shall aim to keep technical details to a minimum, and focus on the choice of quaternion algebra and parameters. This elucidates the connection with the LPS construction. Recall that the meaning of the parameters is similar in both cases: the resulting graphs are $(l+1)$-regular and their size depends on the value of p. Varying p (subject to some constraints) produces an infinite family of $(l+1)$-regular Ramanujan graphs. However, we shall see that the constraints imposed on the parameters $\{p, l\}$ by the LPS and Pizer constructions do *not* agree. In Section 8.2 we give an explicit comparison between the admissible values of the parameter p in the example when $l = 5$.

First we wish to summarize the construction via Pizer [20]. In particular we wish to explain the elements of [20, Theorem 5.1]. Details are kept to a minimum; the reader is encouraged to consult *op.cit.* for details, in particular [20, 4]. We mention one feature of Pizer's approach in advance: we shall see that here the graph is given via its adjacency matrix. Note that this is of a different flavor from the LPS case. There the edges of the graph were specified "locally": given a vertex of the graph (as an element of a group in Section 6.1 or as a class of lattices in Section 6.2), its neighbors were specified directly. (See Section 6.4 for an explicit parametrization of the edges at a vertex.) In Pizer's approach the adjacency matrix, a Brandt matrix (associated to an Eichler order in the quaternion algebra) specifies the edge structure of the graph.

8.1 Overview of the Construction

Let us fix $B = B_{p,\infty}$ to be the quaternion algebra over $\mathbb{Q}$ that is ramified precisely at p and at infinity. We shall consider orders $\mathcal{O}$ of level $N = pM$ and $N = p^2M$ in B, where M is coprime to p. The vertex set of our graph $G(N, l)$ shall be in bijection with (a subset of) the class group of $\mathcal{O}$. The class number of $\mathcal{O}$ depends only on the level of the order and hence we may write $H(pM)$ or $H(p^2M)$ for the size of such a graph. In the case where $M = 1$ by the Eichler class number formula [20, Proposition 4.4] we have

$$H(p) = \frac{p-1}{12} + \frac{1}{4}\left(1 - \left(\frac{-4}{p}\right)\right) + \frac{1}{3}\left(1 - \left(\frac{-3}{p}\right)\right); \tag{35}$$

$$H(p^2) = \frac{p^2-1}{12} + \begin{cases} 0 & \text{if } p \geq 5 \\ \frac{4}{3} & \text{if } p = 3 \end{cases} \tag{36}$$

where $\left(\frac{\cdot}{\cdot}\right)$ is the Kronecker symbol.

The vertex set of $G(N, l)$ shall have $H(N)$ elements when $N = pM$ and when $N = p^2M$ and l is a quadratic nonresidue modulo p. (Note that in this case the graph $G(p^2M, l)$ is bipartite.) For $N = p^2M$ and l a quadratic residue modulo p the graph $G(p^2M, l)$ is non-bipartite of size $\frac{H(p^2M)}{2}$. Recall that a similar dichotomy (between bipartite and non-bipartite cases) exists in the LPS construction as well. The following table summarizes the size of $G(p, l)$ and $G(p^2, l)$ for the case where $\left(\frac{l}{p}\right) = 1$ (and $p > 3$).

$p \mod 12$	$H(p)$	$\frac{H(p^2)}{2}$
1	$\frac{p-1}{12}$	
5	$\frac{p+7}{12}$	$\frac{p^2-1}{12}$
7	$\frac{p+5}{12}$	
11	$\frac{p+13}{12}$	

$$\tag{37}$$

The edge structure of the graph $G(N, l)$ is determined via the adjacency matrix. Recall that the rows and columns of the adjacency matrix of a graph are indexed by the vertex set. One entry of the matrix determines the number of edges between the vertices corresponding to its indices. The edge structure of $G(N, l)$ is given by a *Brandt matrix*. There is a space of modular forms associated to the order $\mathcal{O}$ of the quaternion algebra. This space has dimension as in (37) and it carries the action of a Hecke algebra. For every integer l (coprime to p) the Brandt matrix $B(N, l)$ describes the explicit action of a particular Hecke operator (T_l) on this space.

Restrictions on the parameters p and l guarantee that $B(N, l)$ is in fact the adjacency matrix of a graph. Properties of the resulting graph (e.g., the graph being simple and connected, as well as statements about its spectrum and girth) can be phrased as statements about the Brandt matrices $B(N, l)$ and in turn studied as statements about modular forms.

To ensure the edges of the graph $G(N, l)$ are undirected, $B(N, l)$ must be symmetric. By [20, Proposition 4.6] this is the case for $N = pM$ if $p \equiv 1 \mod 12$ and for $N = p^2 M$ if $p > 3$.

To ensure the graph has no loops we must have $\text{tr} B(N, l) = 0$, and for no multiple edges $\text{tr}(B(N, l))^2 = 0$. By [20, Proposition 4.8] these translate to the conditions $\text{tr} B(N, l) = 0$, $\text{tr} B(N, l^2) = H(N)$. (This depends on the relationship of the traces within a family of Brandt matrices $B(N, l)$ for fixed N and varying l.) These traces can be given in terms of parameters dependent on the order $\mathcal{O}$ [20, Proposition 4.9].

It turns out that the above conditions together already guarantee that $B(N, l)$ determines a Ramanujan graph. This is the content of the following theorem.

Theorem 8.1 ([20, Theorem 5.1]). *Let l be a prime coprime to pM and let $N = pM$. Consider the graph $G(N, l)$ determined by the Brandt matrix $B(N, l)$ as its adjacency matrix. Assume that $B(N, l)$ is symmetric, $\text{tr} B(N, l) = 0$ and $\text{tr} B(N, l^2) = H(N)$. Then $G(N, l)$ is a non-bipartite $(l + 1)$-regular simple Ramanujan graph on $H(N)$ vertices.*

Similarly, let $N = p^2 M$ and assume the above conditions $\text{tr} B(N, l) = 0$ and $\text{tr} B(N, l^2) = H(N)$ hold. If l is a quadratic nonresidue modulo p, then $B(N, l)$ is the adjacency matrix of a bipartite $(l + 1)$-regular simple Ramanujan graph on $H(N)$ vertices. If l is a quadratic residue modulo p, then $B(N, l)$ is the adjacency matrix of two copies of an $(l + 1)$-regular simple non-bipartite Ramanujan graph on $\frac{H(N)}{2}$ vertices.

Recall that the quaternion algebra B underlying the construction above is ramified at exactly two places, p and ∞. This uniquely determines the algebra $B = B_{p,\infty}$ (cf. [20, Proposition 4.1]). Given a specific l one may ask for what p primes and $N = p$ are the conditions $\text{tr} B(N, l) = 0$ and $\text{tr} B(N, l^2) = H(N)$ satisfied. This can be answered by translating the conditions to modular conditions on p. This is carried out for $l = 2$ in [20, Example 2]. In the LPS construction above we were interested in $l + 1$ regular graphs where $l \equiv 1 \mod 4$. To compare the families of Ramanujan graphs emerging from the two constructions, in the next section we carry out the same computation for $l = 5$.

8.2 The Size of a Six-Regular Pizer Graph

We wish to consider a special case of Pizer's construction in [20, Section 5] where the order $\mathcal{O}$ is a (level p) maximal order in $B_{p,\infty}$ and the Ramanujan graph is $l + 1$ regular. In particular, we are interested in the case where $l = 5$. (Since the LPS construction discussed in Section 6 requires $l \equiv 1 \mod 4$, this is the smallest l where a comparison can be made.) In this section we follow the methods of [20, Example 2] to give explicit modular conditions on p to satisfy Pizer's construction. The Brandt matrix $B(p; 5)$ associated to the maximal order $\mathcal{O} \subset B_{p,\infty}$ (of level p) is a square matrix of size $H(p)$. It follows from Theorem 8.1 [20, Proposition 5.1] that it is the adjacency matrix of a 6-regular simple Ramanujan graph if the following conditions hold:

1. $p \equiv 1 \mod 12$
2. $\operatorname{tr} B(p, 5) = 0$
3. $\operatorname{tr} B(p, 5^2) = \operatorname{Cl} \mathcal{O}$

Note that here Condition 1 guarantees that the graph is symmetric, and Condition 2 that it has no loops. By [20, Proposition 4.4] the condition $p \equiv 1 \mod 12$ gives $\operatorname{Cl}(\mathcal{O}) = \operatorname{Mass} \mathcal{O} = \frac{p-1}{12}$.

The Conditions 2 and 3 concern the trace of the Brandt matrices $B(p, 5)$ and $B(p, 25)$ associated to $\mathcal{O}$ of level p. These can be computed using [20, Proposition 4.9]. In particular, *loc. cit.* guarantees that Conditions 2 and 3 hold under certain conditions. To state these conditions we must introduce some notation. For $m = 5$ and $m = 25$, respectively, let s be an integer such that $\Delta = s^2 - 4m$ is negative. Let t and r be chosen such that

$$\Delta = s^2 - 4 \cdot m = \begin{cases} t^2 r & 0 > r \equiv 1 \mod 4 \\ t^2 4r & 0 > r \equiv 2, 3 \mod 4 \end{cases} \tag{38}$$

Let f be any positive divisor of t and $d := \frac{\Delta}{f^2}$. Let $c(s, f, p)$ denote the number of embeddings of $\mathcal{O}_p^d$ into $\mathcal{O}_p$ that are inequivalent modulo the unit group $U(\mathcal{O}_p)$. By [20, Proposition 4.9] we have that

$$\text{Condition 2 is satisfied} \iff c(s, f, p) = 0 \text{ for every } s, f \text{ with } m = 5 \tag{39}$$

$$\text{Condition 3 is satisfied} \iff c(s, f, p) = 0 \text{ for every } s, f \text{ with } m = 5^2 \tag{40}$$

The integers $c(s, f, p)$ are given in tables in [18, pp. 692–693]. We use information in these tables to translate the conditions (39) and (40) into modular conditions on p.

First, if $m = 5$, the possible values of $s, \Delta, r, t,$ and f are as follows:

s	0	1	2		3	4
Δ	−20	−19	−16		−11	−4
t	1	1	2		1	1
r	−5	−19	−1		−11	−1
f	1	1	1	2	1	1
d	−20	−19	−16	−4	−11	−4

It follows from Condition 1 that $p \nmid d = \frac{\Delta}{f^2}$. It follows from the tables in [18, pp. 692–693] that $c(s, f, p) = c(s, f, p)_{p^{2 \cdot 0 + 1}} = 0$ if and only if d is the square of a unit in $\mathbb{Z}_p$, i.e., a quadratic residue modulo p. By Condition 1 we certainly have $\left(\frac{-4}{p}\right) = \left(\frac{-16}{p}\right) = 1$ and by quadratic reciprocity $\left(\frac{d}{p}\right) = 1$ is equivalent to $\left(\frac{p}{d}\right) = 1$. It follows that by (39) that Condition 2 is satisfied if in addition to Condition 1 p satisfies the following modular conditions.

$c(s, f, p)$	$\Delta = d$	condition	
$c(0, 1, p)$	−20	$p \in \{1, 4\} \mod 5$	(41)
$c(1, 1, p)$	−19	$p \in \{1, 4, 5, 6, 7, 9, 11, 16, 17\} \mod 19$	
$c(3, 1, p)$	−11	$p \in \{1, 3, 4, 5, 9\} \mod 11$	

Second, to guarantee that the conditions in (40) are satisfied, let $m = 25$. Then the possible values of s, Δ, r, t and f are as follows:

s	0	1	2	3	4	5	6	7	8	9	
Δ	−100	−99	−96	−91	−84	−75	−64	−51	−36	−19	(42)
t	5	3	4	1	1	5	4	1	3	1	
r	−1	−11	−6	−91	−21	−3	−1	−51	−1	−19	
f	1, 5	1, 3	1, 2, 4	1	1	1, 5	1, 2, 4	1	1, 3	1	

By (1) and (41) we have that $p \nmid d$ for any of the above values of Δ and $d = \frac{\Delta}{f^2}$. Then it again follows from the tables in [18, pp. 692–693] that (40) is satisfied if and only if for any such d $\left(\frac{d}{p}\right) = 1$ or, equivalently by (1), $\left(\frac{p}{d}\right) = 1$. By properties of the Legendre symbol and the previously imposed conditions on the residue class of p modulo 12, 5, 11, and 19 this is true for $\Delta \in \{-100, -99, -75, -64, -36, -19\}$. The remaining cases amount to the following additional modular conditions on p :

Δ	$d = \frac{\Delta}{f^2}$	condition	
−96	−96, −24 or − 6	$p \in \{1, 7\} \mod 8$	
−51	$-51 = -3 \cdot 17$	$p \in \{1, 2, 4, 8, 9, 13, 15, 16\} \mod 17$	(43)
−84	$-84 = -12 \cdot 7$	$p \in \{1, 2, 4\} \mod 7$	
−91	$-91 = -7 \cdot 13$	$p \in \{1, 3, 4, 9, 10, 12\} \mod 13$	

We summarize the modular conditions on p in the following corollary.

Corollary 8.2. *The Brandt matrix $B(p; 5)$ associated to a maximal order in $B_{p,\infty}$ by Pizer [20] is the adjacency matrix of a 6-regular simple, connected, non-bipartite Ramanujan graph if and only if p satisfies the following congruence conditions:*

Modulus	Remainders allowed
24	1
5	1, 4
7	1, 2, 4
11	1, 3, 4, 5, 9
13	1, 3, 4, 9, 10, 12
17	1, 2, 4, 8, 9, 13, 15, 16
19	1, 4, 5, 6, 7, 9, 11, 16, 17

$$(44)$$

These conditions are equivalent to saying that $p \equiv 1 \mod 24$ and p is a quadratic residue modulo the primes $5, 7, 11, 13, 17, 19$. Note that p may belong to one of $1 \cdot 2 \cdot 3 \cdot 5 \cdot 6 \cdot 8 \cdot 9 = 12\,960$ residue classes modulo $24 \cdot 5 \cdot 7 \cdot 11 \cdot 13 \cdot 17 \cdot 19 = 38\,798\,760$.

The corollary describes the set of primes p for which $G(p, 5)$ is a six-regular Ramanujan graph. The condition $p \equiv 1 \mod 4$, $p \equiv 1, 4 \mod 5 = l$ guarantees that for these primes the LPS construction is a six-regular graph as well.

Remark 8.3. The smallest prime satisfying all the congruence conditions of Corollary 8.2 is 53881. This corresponds to a 6-regular Pizer graph with 4490 vertices. Among the first one million primes, 1670 satisfy all these congruence conditions.

9 Relationship Between LPS and Pizer Constructions

We wish to compare the two different approaches to constructing Ramanujan graphs that we have discussed. Throughout the previous sections, we have seen that the constructions of LPS and Pizer (recall the latter agree with supersingular isogeny graphs for particular choices) have similar elements. In this section, we wish to further highlight these similarities, as well as the discrepancies between the two approaches.

First let us revisit the chains of graph isomorphisms/bijections that the respective constructions fit into. These are as follows:

$$\textbf{(LPS)}\,\mathrm{Cay}(\mathrm{PSL}_2(\mathbb{F}_p), S) \cong \Gamma(2p)\backslash\mathrm{PGL}_2(\mathbb{Q}_l)/\mathrm{PGL}_2(\mathbb{Z}_l)$$

$$\cong G'(\mathbb{Q})\backslash H_{2p}(\mathbb{A}_f)/K_0^{2p}\ (\mathcal{O}[l^{-1}])^\times\backslash\mathrm{GL}_2(\mathbb{Q}_l)/\mathrm{GL}_2(\mathbb{Z}_l)$$

$$\cong B^\times(\mathbb{Q})\backslash B^\times(\mathbb{A}_f)/B^\times(\hat{\mathbb{Z}}) \cong \mathrm{Cl}\mathcal{O} \cong \mathrm{SSIG}\ \textbf{(Pizer)}$$

Recall that in the first line, we have the LPS construction in terms of a Cayley graph on the group $\mathrm{PSL}_2(\mathbb{F}_p)$; it corresponds to the "local double coset graph" defined by taking a finite quotient of an infinite tree of homothety classes of lattices. The vertex set of this graph is in bijection with the adelic double cosets on the right-hand side. (For the sake of this comparison we omitted the infinite place.)

On the right-hand end of the second line, we have the supersingular isogeny graphs discussed in Part 1. These are symmetric simple graphs isomorphic to $G(p, l)$ constructed by Pizer (see Section 8) when $p \equiv 1 \mod 12$. The vertex set of $G(p, l)$ is the class group of a maximal order $\mathcal{O}$ in the quaternion algebra $B_{p,\infty}$. This set is in bijection with the adelic double cosets. Via strong approximation (see Section 7.3) these adelic double cosets are in bijection with local double cosets, which at a place l where $B_{p,\infty}$ splits can be written as the left-hand side object.

Despite the similarities between these chains of bijections, there are significant discrepancies between the two objects. First of all, there is a discrepancy in the underlying quaternion algebras. For the LPS graphs we considered the underlying algebra of Hamiltonian quaternions $(B_{2,\infty})$. Varying the parameter p we get different Ramanujan graphs by changing the congruence subgroup $\Gamma(2p)$ without ever changing the underlying algebra. On the other hand, the Pizer graphs were constructed using $B = B_{p,\infty}$. The underlying quaternion algebra varies with the choice of the parameter p. We note that the construction in LPS can be carried out for any B ramified at ∞ and split at l, and would still result in Ramanujan graphs (see [15, Theorem 7.3.12]). However, in this more general case we do not have a clear path for obtaining an explicit description of these graphs as Cayley graphs. For additional details see [15, Remark 7.4.4(iv)]. If one took $B_{p,\infty}$ for both the LPS and Pizer cases, the infinite families of Ramanujan graphs formed would differ because the LPS family is formed by varying the subgroup $\Gamma(2p)$ (or more generally $\Gamma(N)$ for l a quadratic residue mod N) while the Pizer family is formed by varying the quaternion algebra $B_{p,\infty}$.

Let us consider the choice of parameters next. For the LPS graphs we required only that $l \equiv 1 \mod 4$ and that p is odd and prime to l. If -1 is a quadratic residue modulo p, then the resulting graph is isomorphic to a subgroup of $\mathrm{PGL}_2(\mathbb{Z}/p\mathbb{Z})$ [15, Theorem 7.4.3]. Furthermore, if l is a quadratic residue modulo $2p$, then this graph is non-bipartite and isomorphic to the Cayley graph of $\mathrm{PSL}_2(\mathbb{F}_p)$ with $\frac{p^3-p}{2}$ elements.

In the case of the Pizer graphs $G(N, l)$ we must have $N = pM$ coprime to l. Further congruence conditions on N guarantee properties of the resulting graph (see Section 8), e.g., $p \equiv 1 \mod 12$ guarantees that the adjacency matrix is symmetric. The number of vertices in $G(N, l)$ is then $H(N)$, the class number of an order of level N in $B_{p,\infty}$. For example, if $N = p \equiv 1 \mod 12$, then this results in a graph of size $\frac{p-1}{2}$.

To compare the two in the simplest case when $l \equiv 1 \mod 4$, i.e., $l = 5$, recall that Corollary 8.2 gives the exact congruence conditions on p so that the Pizer construction of the graph $G(p, 5)$ is a six-regular Ramanujan graph on $\frac{p-1}{12}$ vertices. For these primes, the LPS construction also produces a Ramanujan graph. The size of the two graphs is very different. Notice however that when both graphs exist the size of the LPS graph is divisible by the size of the Pizer graph (cf. Remark 8.3).

Let us turn our attention to the local double coset objects in the above chain of bijections. In the second line, corresponding to Pizer graphs, we have $(\mathcal{O}[l^{-1}])^\times$ appearing where $\mathcal{O}$ is an order of the quaternion algebra $B_{p,\infty}$. For the graph $G(p, l)$ this $\mathcal{O}$ is an order of level p, i.e., a maximal order. The corresponding subgroup $(\mathcal{O}[l^{-1}])^\times$ of $B^\times(\mathbb{Z}[l^{-1}])$ is analogous to the subgroup $\Gamma = G'(\mathbb{Z}[l^{-1}])$ for the LPS construction. This is much larger than the congruence subgroup $\Gamma(2p) \leq \Gamma$ that appears in the local double coset objects in that case.

The fact that the LPS construction involves this *smaller* congruence subgroup $\Gamma(2p)$ also accounts for the discrepancy between the two lines at the adelic double cosets. Recall from Section 7.2 that H_{2p} was not the entire $G'(\mathbb{A})$ but instead a finite index normal subgroup of it. We note that if one replaced $\Gamma(2p)$ in the LPS construction with $\Gamma(2N)$, where $p \mid N$, the LPS graph $\Gamma(2N)\backslash\mathrm{PGL}_2(\mathbb{Q}_l)/\mathrm{PGL}_2(\mathbb{Z}_l)$ is a finite cover of $\Gamma(2p)\backslash\mathrm{PGL}_2(\mathbb{Q}_l)/\mathrm{PGL}_2(\mathbb{Z}_l)$ [12, Section 3].

One may wonder if an object analogous to $\mathrm{Cl}(\mathcal{O})$ could be appended to the chain of bijections for LPS graphs. Or even if, in the local double coset object for LPS graphs $\Gamma(2p)$ could be written as $(\mathcal{O}_{2p}(\mathbb{Z}[l^{-1}]))^\times$ as well, for a quaternion order $\mathcal{O}_{2p}$. (More precisely, if $\Gamma(2p)$ agrees with the image of $(\mathcal{O}_{2p}(\mathbb{Z}[l^{-1}]))^\times$ under the map $B^\times \to G'$ for some order $\mathcal{O}_{2p}$.)

The answer to the second question is affirmative. Using the basis $1, \mathbf{i}, \mathbf{j}, \mathbf{k}$ for $B = B_{2,\infty}$ the requisite relationship holds between $\mathcal{O}_{2p}$ and $\Gamma(2p)$ for the order $\mathcal{O}_{2p}$ spanned by $\{1, 2p\mathbf{i}, 2p\mathbf{j}, 2p\mathbf{k}\}$. Note that this order has level 2^5p^3, hence it is not an Eichler order.

We remark that the size of the class set of this $\mathcal{O}_{2p}$ can be computed using [19, Theorem 1.12] and it turns out to be $\frac{4p^2(p+1)+4}{3}$ or $\frac{4p^2(p+1)}{3}$ if $p \equiv 1 \mod 3$ or $p \equiv 2 \mod 3$, respectively. This is clearly different from the size of $\mathrm{PSL}_2(\mathbb{F}_p)$ which is a numerical obstruction to extending the chain of isomorphisms for LPS graphs analogously to the row for Pizer graphs.

References

1. Gora Adj, Omran Ahmadi, and Alfred Menezes, *On isogeny graphs of supersingular elliptic curves over finite fields*, Cryptology ePrint Archive, Report 2018/132, 2018, https://eprint.iacr.org/2018/132.

2. Noga Alon, *Eigenvalues and expanders*, Combinatorica **6** (1986), no. 2, 83–96, Theory of computing (Singer Island, Fla., 1984). MR 875835

3. Denis X. Charles, Eyal Z. Goren, and Kristin E. Lauter, *Cryptographic hash functions from expander graphs*, J. Cryptology **22** (2009), no. 1, 93–113, available at https://eprint.iacr.org/2006/021.pdf. MR 2496385

4. _____, *Families of Ramanujan graphs and quaternion algebras*, Groups and symmetries, CRM Proc. Lecture Notes, vol. 47, Amer. Math. Soc., Providence, RI, 2009, pp. 53–80. MR 2500554

5. Gaëtan Chenevier, *Lecture notes*, 2010, http://gaetan.chenevier.perso.math.cnrs.fr/coursIHP/chenevier_lecture6.pdf, retrieved August 13, 2017.

6. Pierre Deligne, *Formes modulaires et représentations l-adiques*, Séminaire Bourbaki. Vol. 1968/69, vol. 179, Lecture Notes in Math., no. 355, Springer, Berlin, 1971, pp. 139–172.

7. _____, *La conjecture de Weil. I*, Publications Mathématiques de l'Institut des Hautes Études Scientifiques **43** (1974), no. 1, 273–307.

8. Luca De Feo, David Jao, and Jérôme Plût, *Towards quantum-resistant cryptosystems from supersingular elliptic curve isogenies*, J. Math. Cryptol. **8** (2014), no. 3, 209–247. MR 3259113

9. Stephen S. Gelbart, *Automorphic forms on adele groups*, no. 83, Princeton University Press, 1975.

10. Yasutaka Ihara, *Discrete subgroups of* PL$(2, k_\wp)$, Algebraic Groups and Discontinuous Subgroups (Proc. Sympos. Pure Math., Boulder, Colo., 1965), Amer. Math. Soc., Providence, R.I., 1966, pp. 272–278. MR 0205952

11. David Jao, Stephen D Miller, and Ramarathnam Venkatesan, *Do all elliptic curves of the same order have the same difficulty of discrete log?*, International Conference on the Theory and Application of Cryptology and Information Security, Springer, 2005, pp. 21–40.

12. Wen-Ch'ing Winnie Li, *A survey of Ramanujan graphs*, Arithmetic, geometry and coding theory (Luminy, 1993), de Gruyter, Berlin, 1996, pp. 127–143. MR 1394930

13. Eyal Lubetzky and Yuval Peres, *Cutoff on all Ramanujan graphs*, Geometric and Functional Analysis **26** (2016), no. 4, 1190–1216.

14. Alexander Lubotzky, Richard L. Phillips, and Peter Sarnak, *Ramanujan graphs*, Combinatorica **8** (1988), no. 3, 261–277. MR 963118 (89m:05099)

15. Alexander Lubotzky, *Discrete groups, expanding graphs and invariant measures*, Modern Birkhäuser Classics, Birkhäuser Verlag, Basel, 2010, With an appendix by Jonathan D. Rogawski, Reprint of the 1994 edition. MR 2569682

16. Jean-Francois Mestre, *La méthode des graphes. Exemples et applications*, Proceedings of the International Conference on Class Numbers and Fundamental Units of Algebraic Number Fields (Katata, 1986), Nagoya Univ., Nagoya, 1986, pp. 217–242. MR 891898

17. Christophe Petit, Kristin Lauter, and Jean-Jacques Quisquater, *Full cryptanalysis of LPS and Morgenstern hash functions*, Security and Cryptography for Networks (Berlin, Heidelberg) (Rafail Ostrovsky, Roberto De Prisco, and Ivan Visconti, eds.), Springer Berlin Heidelberg, 2008, pp. 263–277.

18. Arnold Pizer, *The representability of modular forms by theta series*, Journal of the Mathematical Society of Japan **28** (1976), no. 4, 689–698.

19. _____, *An algorithm for computing modular forms on* $\Gamma_0(N)$, Journal of Algebra **64** (1980), no. 2, 340–390.

20. _____, *Ramanujan graphs*, Computational perspectives on number theory (Chicago, IL, 1995), AMS/IP Stud. Adv. Math., vol. 7, Amer. Math. Soc., Providence, RI, 1998, pp. 159–178. MR 1486836

21. *Post-Quantum Cryptography Standardization*, https://csrc.nist.gov/Projects/Post-Quantum-Cryptography/Post-Quantum-Cryptography-Standardization, Accessed: 2018-04-14.

22. Naser T. Sardari, *Diameter of Ramanujan graphs and random Cayley graphs*, (2018). Combinatorica, 1–20. https://doi.org/10.1007/s00493-017-3605-0

23. Joseph H. Silverman, *The arithmetic of elliptic curves*, second ed., Graduate Texts in Mathematics, vol. 106, Springer, Berlin–Heidelberg–New York, 2009.

24. Jean-Pierre Tillich and Gilles Zémor, *Collisions for the LPS expander graph hash function*, Advances in Cryptology – EUROCRYPT 2008 (Nigel Smart, ed.), Springer, 2008, pp. 254–269.

25. Jacques Vélu, *Isogénies entre courbes elliptiques*, C. R. Acad. Sci. Paris Sér. A-B **273** (1971), A238–A241. MR 0294345

26. Marie-France Vignéras, *Arithmétique des algèbres de quaternions*, Lecture Notes in Mathematics, vol. 800, Springer, Berlin, 1980. MR 580949

27. John Voight, *Quaternion algebras*, 2018, https://math.dartmouth.edu/~jvoight/quat-book.pdf, retrieved October 20, 2017.

Cycles in the Supersingular ℓ-Isogeny Graph and Corresponding Endomorphisms

Efrat Bank, Catalina Camacho-Navarro, Kirsten Eisenträger, Travis Morrison, and Jennifer Park

Abstract We study the problem of generating the endomorphism ring of a supersingular elliptic curve by two cycles in ℓ-isogeny graphs. We prove a necessary and sufficient condition for the two endomorphisms corresponding to two cycles to be linearly independent, expanding on the work by Kohel in his thesis. We also give a criterion under which the ring generated by two cycles is not a maximal order. We give some examples in which we compute cycles which generate the full endomorphism ring. The most difficult part of these computations is the calculation of the trace of these cycles. We show that a generalization of Schoof's algorithm can accomplish this computation efficiently.

Keywords Supersingular elliptic curves · Endomorphism rings · Isogenies · Isogeny graphs · Quaternion algebras

1 Introduction

The currently used cryptosystems, such as RSA and systems based on elliptic curve cryptography (ECC), are known to be broken by quantum computers. However,

E. Bank · J. Park
Department of Mathematics, University of Michigan, Ann Arbor, MI, USA
e-mail: ebank@umich.edu; jmypark@umich.edu; http://www-personal.umich.edu/~ebank/;
http://www-personal.umich.edu/~jmypark/

C. Camacho-Navarro
Department of Mathematics, Colorado State University, Fort Collins, CO 80523, USA
e-mail: camacho@math.colostate.edu

K. Eisenträger
Department of Mathematics, The Pennsylvania State University, University Park, PA 16802, USA
e-mail: eisentra@math.psu.edu; http://www.personal.psu.edu/kxe8/

T. Morrison (✉)
Institute for Quantum Computing, The University of Waterloo, Waterloo, ON, Canada
e-mail: travis.morrison@uwaterloo.ca

© The Author(s) and The Association for Women in Mathematics 2019
J. S. Balakrishnan et al. (eds.), *Research Directions in Number Theory*, Association for Women in Mathematics Series 19, https://doi.org/10.1007/978-3-030-19478-9_2

it is not known whether cryptosystems based on the hardness of computing endomorphism rings or isogenies between supersingular elliptic curves can be broken by quantum computers. Because of this, these systems have been studied intensely over the last few years, and the International Post-Quantum Cryptography Competition sponsored by NIST [18] has further increased interest in studying the security of these systems. There is a submission under consideration [1] which is based on supersingular isogenies.

Cryptographic applications based on the hardness of computing isogenies between supersingular elliptic curves were first given in [6]. In this paper, Charles, Goren, and Lauter constructed a hash function from the ℓ-isogeny graph of supersingular elliptic curves, and finding preimages for the hash function is connected to finding certain ℓ-power isogenies (for a small prime ℓ) between supersingular elliptic curves.

More recently, De Feo, Jao, and Plût [8] proposed post-quantum key-exchange and encryption schemes based on computing isogenies of supersingular elliptic curves. A signature scheme based on supersingular isogenies is given in [28], and [11] gives a signature scheme in which the secret key is a maximal order isomorphic to the endomorphism ring of a supersingular elliptic curve.

There are currently no subexponential classical or quantum attacks for these systems. However, under some heuristic assumptions, the quaternion analogue for the underlying hardness assumption of the hash function in [6] was broken in [12], which suggests that a careful study of the isogenies and endomorphism rings of supersingular elliptic curves is necessary.

For a fixed q, [4, 14] list all isomorphism classes of supersingular elliptic curves over $\mathbb{F}_q$ along with their maximal orders in a quaternion algebra (which was improved in [5, §5.2]). In [15] McMurdy also computes explicit endomorphism rings for some supersingular elliptic curves. The problem of computing isogenies between supersingular elliptic curves over p has been studied, both in the classical setting [9, Section 4] where the complexity of the algorithm is $\tilde{O}(p^{1/2})$, and in the quantum setting [2], where the complexity is $\tilde{O}(p^{1/4})$. In fact, computing the endomorphism ring of a supersingular elliptic curve is deeply connected to computing isogenies between supersingular elliptic curves, as shown by [13]. Heuristic arguments show that these two problems are equivalent [10–12].

In this paper, we work over finite fields $\mathbb{F}_q$ of characteristic p, and study the problem of generating the endomorphism ring of a supersingular elliptic curve E by two cycles in the ℓ-isogeny graph (Definition 2.1) of supersingular elliptic curves. Computing the endomorphism ring of a supersingular elliptic curve via ℓ-isogeny graphs was first studied by Kohel [13, Theorem 75], who gave an approach for finding four linearly independent endomorphisms that generate a finite-index suborder of $\text{End}(E)$ by finding cycles in the ℓ-isogeny graph. The running time of the probabilistic algorithm is $O(p^{1+\varepsilon})$. We demonstrate some obstructions to generating the full endomorphism ring with two cycles $\alpha, \beta \in \text{End}(E)$.

Expanding on [13], we prove in Theorem 4.10 the necessary and sufficient conditions for α and β to be linearly independent. We also prove sufficient conditions for when α and β generate a proper suborder of $\text{End}(E)$ in Theorem 5.1, then compute some examples. In order to do this, we need to detect when the order generated by two cycles is isomorphic to another given order; in §3, we give a criterion that reduces this problem to computing traces of various endomorphisms. In the Appendix we give a generalization of Schoof's algorithm [20] (using the improvements from [23]) and show that the trace of an arbitrary endomorphism of norm ℓ^e can be computed in time polynomial in e, ℓ, and $\log p$.

The paper is organized as follows: In §2 we review some definitions about elliptic curves and define isogeny graphs. In §3 we discuss some background on quaternion algebras, Deuring's correspondence, and we discuss how to compute the endomorphism ring of a supersingular elliptic curve from cycles in the supersingular ℓ-isogeny graph. In §4 we give a necessary and sufficient condition for two endomorphisms to be linearly independent, expanding on a result by Kohel [13]. In §5 we give conditions under which two cycles α, β in the supersingular ℓ-isogeny graph generate a proper suborder of the endomorphism ring. In §6 we compute some examples, and in the Appendix we give the generalization of Schoof's algorithm.

Acknowledgments. We are deeply grateful to John Voight for several helpful discussions, and for providing us with some code that became a part of the code that generated the computational examples in Section 6. We also thank Sean Hallgren and Rachel Pries for helpful discussions. We thank Andrew Sutherland for many comments on an earlier version that led to a significant improvement in the running time analysis of the generalization of Schoof's algorithm, and for outlining the proof of Proposition 2.4. Finally, we thank the Women in Numbers 4 conference and BIRS, for enabling us to start this project in a productive environment. K.E. was partially supported by National Science Foundation awards DMS-1056703 and CNS-1617802. T.M. was partially supported by National Science Foundation awards DMS-1056703 and CNS-1617802, and by funding from the Natural Sciences and Engineering Research Council of Canada, the Canada First Research Excellence Fund, CryptoWorks21, Public Works and Government Services Canada, and the Royal Bank of Canada. C.CN. was partially supported by Universidad de Costa Rica.

2 Isogeny Graphs

2.1 Definitions and Properties

In this section, we recall several definitions and notation that are used throughout. We refer the reader to [22] and [13] for a detailed overview on some of the below. Let k be a field of characteristic $p > 3$.

By an elliptic curve E over a field k, we mean the projective curve with an affine model $E : y^2 = x^3 + Ax + B$ for some $A, B \in k$. The points of E are the points (x, y) satisfying the curve equation, together with the point at infinity. These points form an abelian group. The j-invariant of an elliptic curve given as above is

$j(E) = \frac{256 \cdot 27 \cdot A^3}{4A^3 + 27B^2}$. Two elliptic curves E, E' defined over a field k have the same j-invariant if and only if they are isomorphic over the algebraic closure of k.

Let E_1 and E_2 be elliptic curves defined over k. An *isogeny* $\varphi : E_1 \to E_2$ defined over k is a non-constant rational map which is also a group homomorphism from $E_1(k)$ to $E_2(k)$ [22, III.4]. The *degree* of an isogeny is its degree as a rational map. When the degree d of the isogeny φ is coprime to p, then φ is separable and every separable isogeny of degree $d > 1$ can be factored into a composition of isogenies of prime degrees such that the product of the degrees equals d. If $\psi : E_1 \to E_2$ is an isogeny of degree d, the *dual isogeny* of ψ is the unique isogeny $\widehat{\psi} : E_2 \to E_1$ satisfying $\widehat{\psi}\psi = [d]$, where $[d] : E_1 \to E_1$ is the multiplication-by-d map.

In this paper we will be interested in isogenies of ℓ-power degree, for ℓ a prime different from the characteristic of k. We can describe a separable isogeny from an elliptic curve E to some other elliptic curve via its kernel. Given an elliptic curve E and a finite subgroup H of E, there is a separable isogeny $\varphi : E \to E'$ having kernel H which is unique up to isomorphism (see [22, III.4.12]). In this paper we will identify two isogenies if they have the same kernel. We can compute equations for the isogeny from its kernel by using Vélu's formula [25].

An isogeny of an elliptic curve E to itself is called an endomorphism of E. If E is defined over some finite field $\mathbb{F}_q$, then the set of endomorphisms of E defined over $\overline{\mathbb{F}}_q$ together with the zero map form a ring under the operations addition and composition. It is called the *endomorphism ring* of E, and is denoted by $\mathrm{End}(E)$. It is isomorphic either to an order in a quadratic imaginary field or to an order in a quaternion algebra. In the first case we call E an *ordinary elliptic curve*. An elliptic curve whose endomorphism is isomorphic to an order in a quaternion algebra is called a *supersingular elliptic curve*. Every supersingular elliptic curve over a field of characteristic p has a model that is defined over $\mathbb{F}_{p^2}$ because the j-invariant of such a curve is in $\mathbb{F}_{p^2}$. Given $j \in \overline{\mathbb{F}}_q$ such that $j \neq 0, 1728$, we write $E(j)$ for the curve defined by the equation

$$y^2 + xy = x^3 - \frac{36}{j - 1728}x - \frac{1}{j - 1728}. \tag{2.1.1}$$

Such a curve can be put into a short Weierstrass equation $y^2 = x^3 + Ax + B$. We also write $E(0)$ and $E(1728)$ for the curves with equations $y^2 = x^3 + 1$ and $y^2 = x^3 + x$, respectively.

We borrow definitions from [24, 2.3] to define the ℓ-isogeny graph of supersingular elliptic curves in characteristic p. Given a prime ℓ, the *modular polynomial* $\Phi_\ell(X, Y) \in \mathbb{Z}[X, Y]$ has the property that if $j, j' \in \mathbb{F}_q$, then $\Phi_\ell(j, j') = 0$ if and only if there are elliptic curves $E, E'/\mathbb{F}_q$ with j-invariants j and j', respectively, such that there is a separable isogeny $\phi : E \to E'$ of degree ℓ. When $j \neq 0, 1728$, $j, j' \in \mathbb{F}_q$, and E is ordinary, we can choose E' and ϕ such that ϕ is defined over $\mathbb{F}_q$ [21, Proposition 6.1]. We state and prove a similar result for supersingular elliptic curves E in Corollary 2.5.

Definition 2.1. Let ℓ be a prime different from p. The *supersingular ℓ-isogeny graph in characteristic p* is the multigraph $G(p, \ell)$ whose vertex set is

$$V = V(G(p, \ell)) = \{j \in \mathbb{F}_{p^2} : E(j) \text{ is supersingular}\},$$

and the number of directed edges from j to j' is equal to the multiplicity of j' as a root of $\Phi_\ell(j, Y)$. We can identify an edge between $E(j)$ and $E(j')$ with a cyclic isogeny $\phi : E(j) \to E(j')$ of degree ℓ. Such an isogeny is unique up to post-composing with an automorphism of $E(j')$. By the above discussion, we can label the edges of $G(p, \ell)$ which start at a given vertex $E(j)$ with the $\ell + 1$ chosen isogenies whose kernels are the nontrivial cyclic subgroups of $E(j)[\ell]$.

Let E, E' be two supersingular elliptic curves defined over $\mathbb{F}_{p^2}$. For each prime $\ell \neq p$, E and E' are connected by a chain of isogenies of degree ℓ [16]. By [13, Theorem 79], E and E' can be connected by m isogenies of degree ℓ (and hence by a single isogeny of degree ℓ^m), where $m = O(\log p)$. If ℓ is a fixed prime such that $\ell = O(\log p)$, then any ℓ-isogeny in the chain above can either be specified by rational maps or by giving the kernel of the isogeny, and both of these representations have size polynomial in $\log p$.

The theorem below summarizes several properties of the supersingular ℓ-isogeny graph mentioned above.

Theorem 2.2. Let $\ell \neq p$ be a prime, and let $G(p, \ell)$ be the supersingular ℓ-isogeny graph in characteristic p as in Definition 2.1.

(1) G is connected.
(2) G is $(\ell + 1)$-regular as a directed graph.
(3) $\#V = \left[\frac{p}{12}\right] + \varepsilon_p$. Here,

$$\varepsilon_p \overset{\text{def}}{=} \begin{cases} 0, & p \equiv 1 \pmod{12} \\ 1, & p = 3 \\ 1, & p \equiv 5, 7 \pmod{12} \\ 2, & p \equiv 11 \pmod{12}. \end{cases}$$

Remark 2.3. We have an exception to the $\ell + 1$-regularity at the vertices and their neighbors corresponding to elliptic curves with $j = 0, 1728$, due to their extra automorphisms.

As $G(p, \ell)$ is connected, for any two supersingular elliptic curves E, E' defined over $\mathbb{F}_{p^2}$ there is an isogeny $\phi : E \to E'$ of degree ℓ^e for some e. In fact, we can take this isogeny to be defined over an extension of $\mathbb{F}_{p^2}$ of degree at most 6. If $E/\mathbb{F}_q$ is an elliptic curve and $n \geq 1$ is an integer, we denote the ring of endomorphisms of E which are defined over $\mathbb{F}_{q^n}$ by $\text{End}_{\mathbb{F}_{q^n}}(E)$.

Proposition 2.4. *Let $E/\mathbb{F}_{p^2}$ be a supersingular elliptic curve. Then* $\text{End}_{\mathbb{F}_{p^{2d}}}(E) = \text{End}(E)$, *where* $d = 1$ *if* $j(E) \neq 0, 1728$, $d = 1$ *or* $d = 3$ *if* $j(E) = 0$, *and* $d = 1$ *or* $d = 2$ *if* $j(E) = 1728$.

Proof. First, we observe that for a supersingular elliptic curve $E/\mathbb{F}_q$, all endomorphisms of E are defined over $\mathbb{F}_q$ if and only if the Frobenius endomorphism $\pi : E \to E$ is equal to $[p^k]$ or $[-p^k]$ for some k. This follows from Theorem 4.1 of [27], case (2). This is the case when q is an even power of p and the trace of Frobenius is $\pm 2\sqrt{q}$.

Now assume $E/\mathbb{F}_{p^2}$ is a supersingular elliptic curve and let $\pi : E \to E$ be the Frobenius endomorphism of E. Consider the multiplication-by-p map $[p] : E \to E$. By [22, II.2.12], the map $[p]$ factors as

$$[p] = \alpha \circ \pi,$$

where α is an automorphism of E. The automorphism is defined over $\mathbb{F}_{p^2}$, since $[p]$ and π are defined over $\mathbb{F}_{p^2}$. Thus π and α commute. If $j(E) \neq 0, 1728$, then $\text{Aut}(E) = \{[\pm 1]\}$ and thus $[p] = \pm \pi$. By the above observation, we have $\text{End}_{\mathbb{F}_{p^2}}(E) = \text{End}(E)$.

If $j(E) = 0$, then $\alpha^3 = [\pm 1]$. We also have

$$[p^3] = (\alpha \circ \pi)^3 = \alpha^3 \circ \pi^3,$$

so $\pi^3 = [\pm p^3]$. In this case, the Frobenius of the base change of E to $\mathbb{F}_{p^6}$ is π^3, and thus $\text{End}_{\mathbb{F}_{p^6}}(E) = \text{End}(E)$ again by the above observation. The proof of the case $j(E) = 1728$ is similar. □

Corollary 2.5. *If $E_0, E_1/\mathbb{F}_{p^2}$ are supersingular and if $\ell \neq p$, then*

$$\text{Hom}_{\mathbb{F}_{p^{2d}}}(E_0, E_1) = \text{Hom}(E_0, E_1),$$

where $d = 1, 2, 3$, or 6. If $j(E_i) \neq 0, 1728$ for $i = 0, 1$, then we can take $d = 2$.

Proof. First, we claim that for some $d' \in \{1, 2, 3, 6\}$, we have $\#E_0(\mathbb{F}_{p^{2d'}}) = \#E_1(\mathbb{F}_{p^{2d'}})$. This follows from the fact that all supersingular curves $E/\mathbb{F}_{p^2}$ will have the same number of points over an extension of $\mathbb{F}_{p^2}$ of degree $1, 2, 3$, or 6, which we now prove. The trace of Frobenius of a supersingular elliptic curve $E/\mathbb{F}_{p^2}$ is either $0, \pm p$, or $\pm 2p$, again by Proposition 4.1 of [27]. By inspection, we see that if α, β are the roots of the characteristic polynomial of the (p^2-power) Frobenius endomorphism of E, then α and β are either $\pm p$, $\pm p(1 \pm \sqrt{-3})/2$, $\alpha = \beta = p$, or $\alpha = \beta = -p$. Thus for some $d' \in \{1, 2, 3, 6\}$, $\alpha^{d'} = \beta^{d'} = p^{d'}$. By [22, V.2.3.1(a)],

$$\#E(\mathbb{F}_{p^{2d'}}) = p^{2d'} + 1 - \alpha^{d'} - \beta^{d'} = p^{2d'} + 1 - 2p^{d'}.$$

Thus we can choose $d' = 1, 2, 3$, or 6 such that the claim holds for E_0 and E_1.

Now let $\mathbb{F}_{p^{2d''}}/\mathbb{F}_{p^2}$ be an extension such that $\text{End}_{\mathbb{F}_{p^{2d''}}}(E_i) = \text{End}(E_i)$ for $i = 0, 1$. Let $d = \text{lcm}\{d', d''\}$. By Proposition 2.4, $d \in \{1, 2, 3, 6\}$.

If $d > 1$, we base change E_i to $\mathbb{F}_{p^{2d}}$ and denote these curves again by E_i. For each $i = 0, 1$, let π_{E_i} be the p^{2d}-power Frobenius endomorphism for E_i; then, by our choice of d' which divides d, either $\pi_{E_i} = [p^d] : E_i \to E_i$ or $\pi_{E_i} = [-p^d] : E_i \to E_i$ for $i = 0, 1$. Consequently, for any $\phi \in \text{Hom}(E_0, E_1)$ it follows that

$$\phi \circ \pi_{E_0} = \pi_{E_1} \circ \phi.$$

Thus ϕ is defined over $\mathbb{F}_{p^{2d}}$. □

Remark 2.6. If $E_0, E_1/\mathbb{F}_{p^2}$ are supersingular and $j(E_i) \neq 0, 1728$ for $i = 0, 1$, then E_0 and E_1 are connected by a chain of ℓ-isogenies defined over $\mathbb{F}_{p^4}$.

3 Quaternion Algebras, Endomorphism Rings, and Cycles

3.1 Quaternion Algebras

For $a, b \in \mathbb{Q}^\times$, let $H(a, b)$ denote the quaternion algebra over $\mathbb{Q}$ with basis $1, i, j, ij$ such that $i^2 = a$, $j^2 = b$, and $ij = -ji$. That is,

$$H(a, b) = \mathbb{Q} + \mathbb{Q}i + \mathbb{Q}j + \mathbb{Q}ij.$$

Every 4-dimensional central simple algebra over $\mathbb{Q}$ is isomorphic to $H(a, b)$ for some $a, b \in \mathbb{Q}$; for example, see [26, Proposition 7.6.1].

There is a *canonical involution* on $H(a, b)$ which sends an element $\alpha = a_1 + a_2 i + a_3 j + a_4 ij$ to $\bar{\alpha} := a_1 - a_2 i - a_3 j - a_4 ij$. Define the *reduced trace* of an element α as above to be

$$\text{Trd}(\alpha) = \alpha + \bar{\alpha} = 2a_1,$$

and the *reduced norm* to be

$$\text{Nrd}(\alpha) = \alpha\bar{\alpha} = a_1^2 - a a_2^2 - b a_3^2 + ab a_4^2.$$

Definition 3.1. Let B be a quaternion algebra over $\mathbb{Q}$, and let p be a prime or ∞. Let $\mathbb{Q}_p$ be the p-adic rationals if p is finite, and let $\mathbb{Q}_\infty = \mathbb{R}$. We say that B is *split at* p if

$$B \otimes_{\mathbb{Q}} \mathbb{Q}_p \cong M_2(\mathbb{Q}_p),$$

where $M_2(K)$ is the algebra of 2×2 matrices with coefficients in K. Otherwise B is said to be *ramified at* p.

Orders in quaternion algebras appear as endomorphism rings of some elliptic curves ([7]):

Theorem 3.2 (Deuring's Correspondence). *Let E be an elliptic curve over $\mathbb{F}_p$ and suppose that the $\mathbb{Z}$-rank of $\text{End}(E)$ is 4. Then $B := \text{End}(E) \otimes_{\mathbb{Z}} \mathbb{Q}$ is a quaternion algebra ramified exactly at p and ∞, denoted $B_{p,\infty}$, and $\text{End}(E)$ is isomorphic to a maximal order in $B_{p,\infty}$.*

Under this isomorphism, taking the dual isogeny on $\text{End}(E)$ corresponds to the canonical involution in the quaternion algebra, and thus the degrees and traces of endomorphisms correspond to reduced norms and reduced traces of elements in the quaternion algebra.

Lemma 3.3 ([19], Proposition 5.1). *$B_{p,\infty}$ can be explicitly given as*

(i) $B_{p,\infty} = \left(\frac{-1,-1}{\mathbb{Q}} \right)$ *if $p = 2$;*

(ii) $B_{p,\infty} = \left(\frac{-1,-p}{\mathbb{Q}} \right)$ *if $p \equiv 3 \pmod 4$;*

(iii) $B_{p,\infty} = \left(\frac{-2,-p}{\mathbb{Q}} \right)$ *if $p \equiv 5 \pmod 8$; and*

(iv) $B_{p,\infty} = \left(\frac{-p,-q}{\mathbb{Q}} \right)$ *if $p \equiv 1 \mod 8$, where $q \equiv 3 \mod 4$ is a prime such that q is not a square modulo p.*

The quaternion algebra $B_{p,\infty}$ is an inner product space with respect to the bilinear form

$$\langle x, y \rangle = \frac{\text{Nrd}(x+y) - \text{Nrd}(x) - \text{Nrd}(y)}{2}.$$

The basis $\{1, i, j, ij\}$ is an orthogonal basis with respect to this inner product.

3.2 Computing Endomorphism Rings from Cycles in the ℓ-Isogeny Graphs

Suppose we have an order in $\text{End}(E)$ generated by two cycles in $G(p, \ell)$, which we embed as $\mathcal{O} \subseteq B_{p,\infty}$. Suppose we also have another order $\mathcal{O}' \subseteq B_{p,\infty}$. We want to check whether $\mathcal{O} \simeq \mathcal{O}'$ or not.

One can check this via using the fact that two orders are isomorphic if and only if they are conjugate; this follows from the Skolem-Noether theorem, see [26, Lemma 17.7.2], for example. One can do this by showing that as lattices in a quadratic space $\text{End}(E) \otimes \mathbb{Q} \simeq B_{p,\infty}$, the two lattices are isometric under the quadratic form induced by Nrd. Thus, we can check whether $\mathcal{O} \simeq \mathcal{O}'$ by computing Gram matrices for a basis of each, and checking whether the matrices are conjugate by an orthogonal matrix. The following proposition, which is Corollary 4.4 in [17], makes this remark explicit.

Proposition 3.4. *Two orders $\mathcal{O}, \mathcal{O}' \subseteq B_{p,\infty}$ are conjugate if and only if they are isometric as lattices with respect to the inner product induced by* Nrd. *In particular, for $m, n \in \{1, \ldots, 4\}$, let x_m, y_n be elements in the quaternion algebra $B_{p,\infty}$ such that $\mathcal{O}_1 = \langle x_1, x_2, x_3, x_4 \rangle$ and $\mathcal{O}_2 = \langle y_1, y_2, y_3, y_4 \rangle$ are orders in $B_{p,\infty}$. If $\mathrm{Trd}(x_m \overline{x_n}) = \mathrm{Trd}(y_m \overline{y_n})$ for $m, n \in \{1, 2, 3, 4\}$, then $\mathcal{O}_1 \cong \mathcal{O}_2$.*

Proof. The first statement is [17, Corollary 4.4]. The second statement follows then from the first: the map $x_m \to y_n$ extends linearly to an isometry of lattices in $B_{p,\infty}$. This implies that $\mathcal{O}_1$ and $\mathcal{O}_2$ are conjugate in $B_{p,\infty}$ and hence isomorphic as orders. $\square$

Thus if we have two cycles in $G(p, \ell)$ passing through $E(j)$ which correspond to endomorphisms $\alpha, \beta \in \mathrm{End}(E(j))$, we can generate an order

$$\mathcal{O} = \langle 1, \alpha, \beta, \alpha\beta \rangle = \langle x_0, x_1, x_2, x_3 \rangle \subseteq \mathrm{End}(E) \otimes_{\mathbb{Z}} \mathbb{Q}.$$

Also, suppose we have an order $\mathcal{O}' = \langle y_0, y_1, y_2, y_3 \rangle \subseteq B_{p,\infty}$. Then we can check whether $\mathcal{O} \simeq \mathcal{O}'$ by comparing $\mathrm{tr}(x_m \widehat{x_n})$ and $\mathrm{Trd}(y_m \overline{y_n})$. This idea is used in our examples in §6. Additionally, we use this idea in Theorem 5.1 to produce a geometric obstruction to generating $\mathrm{End}(E(j))$ by two cycles in $G(p, \ell)$.

Lemma 3.5. *Let $\{a_1, \ldots, a_e\}$ be a cycle beginning and ending at a vertex $E(j)$. Then the endomorphism of $E(j)$ corresponding to this cycle has degree ℓ^e.*

Proof. Each edge a_k represents an ℓ-isogeny, which has degree ℓ. Composition of N isogenies of degree ℓ results in an isogeny of degree ℓ^N. $\square$

Theorem 3.6. *Let $C = \{a_1, \ldots, a_e\}$ be a cycle in $G(p, \ell)$ beginning and ending at a vertex $E(j)$ corresponding to an endomorphism of $E(j)$. Then the (reduced) trace of C interpreted as an element of $\mathrm{End}(E(j))$ can be computed in time polynomial in ℓ, $\log p$, and e.*

Proof. This is proved in the Appendix. $\square$

In fact, some of the traces can be recognized immediately without resorting to the modification of Schoof's algorithm.

Lemma 3.7. *The cycles corresponding to the multiplication-by-ℓ^n map (n of the ℓ-isogenies followed by their dual isogenies in reverse order) have trace $2\ell^n$. Suppose $\phi : E \to E'$ is an isogeny and $\rho \in \mathrm{End}(E')$. Then $\mathrm{tr}(\hat{\phi} \circ \rho \circ \phi) = \deg(\phi) \cdot \mathrm{tr}(\rho)$.*

Proof. Let $\phi : E \to E'$ be an isogeny of supersingular elliptic curves. By Proposition 3.9 of [27], the map

$$\iota \otimes \mathrm{id} : \mathrm{End}(E') \otimes \mathbb{Q} \to \mathrm{End}(E) \otimes \mathbb{Q}$$

$$\rho \otimes 1 \mapsto \hat{\phi}\rho\phi \otimes \frac{1}{\deg(\phi)}$$

is an isomorphism of quaternion algebras. It follows that $\mathrm{tr}(\hat{\phi}\rho\phi) = \deg(\phi)\,\mathrm{tr}(\rho)$. $\square$

4   A Condition for Linear Independence

In this section, we prove a necessary and sufficient condition for two endomorphisms α and β to be linearly independent. To prove this we need the notion of a *cycle which has no backtracking*. We first show that this notion is equivalent to the corresponding endomorphism being primitive. Then, in Theorem 4.10, we characterize when two cycles with no backtracking are linearly independent. To do this we use the fact that if two endomorphisms are linearly dependent, then they generate a subring of a quadratic imaginary field, and in particular, they must commute. As a corollary, we obtain that two cycles through a vertex $E(j)$ that do not have the same vertex set must be linearly independent.

Definition 4.1. An isogeny $\phi : E \to E'$ is *primitive* if it does not factor through $[n] : E \to E$ for any natural number $n > 1$.

Remark 4.2. An isogeny $\phi : E \to E'$ is primitive if $\ker(\phi)$ does not contain $E[n]$ for any $n > 1$.

Definition 4.3. Suppose a_1, a_2 are edges in $G(p, \ell)$ whose chosen representatives are ℓ-isogenies $\phi : E(j) \to E(j'), \psi : E(j') \to E(j)$. We say that a_2 is dual to a_1 if $\hat{\phi} \in \mathrm{Aut}(E(j))\psi$. A cycle $\{a_1, \ldots, a_e\}$ in $G(p, \ell)$ *has no backtracking* if a_{i+1} is not dual to a_i for $i = 1, \ldots, e - 1$.

Remark 4.4. Let $\{a_1, \ldots, a_e\}$ be a path in $G(p, \ell)$ and let $\phi_i : E(j_i) \to E(j_{i+1})$ be the chosen isogeny representing a_i for $i = 1, \ldots, e$. Suppose that a_{k+1} is dual to a_k for some $1 \le k \le e - 1$. Then we claim that the isogeny

$$\phi = \phi_e \circ \cdots \circ \phi_1$$

will not be primitive. Since a_{k+1} is dual to a_k, there exists $\rho \in \mathrm{Aut}(E(j_k))$ such that $\phi_k = \widehat{\phi_{k+1}}\rho$. Then $\phi_{k+1} \circ \phi_k = [\ell]\rho$, so ϕ factors through $[\ell]$.

Our definition of a cycle with no backtracking is less restrictive than the notion of a simple cycle in [13], which additionally requires that there are no repeated vertices in the cycle. Proposition 82 of [13] shows that simple cycles in $G(p, \ell)$ through $E(j)$ give rise to primitive endomorphisms. We strengthen this result, proving in Lemma 4.6 below that cycles through $E(j)$ with no backtracking correspond exactly to primitive endomorphisms.

Given a path in $G(p, \ell)$ of length e between j and j', there is an isogeny $\phi : E(j) \to E(j')$ of degree ℓ^e obtained by composing the isogenies representing the edges in the path. If this path has no backtracking, the kernel of ϕ is a cyclic subgroup of order ℓ^e in $E(j)[\ell^e]$. Conversely, given an isogeny $\phi : E(j) \to E(j')$ with cyclic kernel of order ℓ^e, there is a corresponding path in $G(p, \ell)$.

Proposition 4.5. *Suppose that* $\phi : E(j) \to E(j')$ *is an isogeny with cyclic kernel of order* ℓ^e. *There is a unique path in* $G(p, \ell)$ *such that the factorization of* ϕ *into* *a chain of ℓ-isogenies corresponds to the edges in the path, and the path has no backtracking.*

Proof. The proof is by induction on e. If $e = 1$, there is a unique edge corresponding to the isogeny $\phi : E(j) \to E(j')$, because each edge starting at $E(j)$ corresponds to a unique cyclic subgroup of $E(j)[\ell]$. Now suppose that the kernel of $\phi : E(j) \to E(j')$ is generated by a point P of $E(j)$ of order ℓ^e. There is an edge in $G(p, \ell)$ from $E(j)$ to another vertex $E(j_1)$ which is labeled by $\phi_1 : E(j) \to E(j_1)$ and whose kernel is $\langle [\ell^{e-1}]P \rangle$. Then because $\phi([\ell^{e-1}]P) = 0$, we have a factorization $\phi := \psi \circ \phi_1$. Then $\psi : E(j_1) \to E(j')$ has degree ℓ^{e-1} and its kernel is cyclic of order ℓ^{e-1}, generated by $\phi_1(P)$. Then there is a path of length $e - 1$ between $E(j_1)$ and $E(j')$ with no backtracking by the inductive hypothesis. By concatenating with the edge corresponding to ϕ_1, we have a path of length e between $E(j)$ and $E(j')$. Note that the first edge in the path for ψ cannot be dual to the edge for ϕ_1, because otherwise $E(j)[\ell] \subseteq \ker \phi$, which is cyclic by assumption. $\qquad\square$

Given a path C in $G(p, \ell)$ starting at $E(j)$, the *isogeny corresponding to C* is the isogeny obtained by composing the isogenies representing the edges along the path. Conversely, given an isogeny $\phi : E(j) \to E(j')$ with cyclic kernel, the *path corresponding to ϕ* is the path constructed as above. We remark that it is the kernel of an isogeny which determines the path in $G(p, \ell)$, so two distinct primitive isogenies will determine the same path if they have the same kernel. This path is only unique because we fix an isogeny representing each edge.

Lemma 4.6. *Let $\{a_1, \ldots, a_e\}$ be a cycle in $G(p, \ell)$ through the vertex $E(j)$ with corresponding endomorphism $\alpha \in \mathrm{End}(E(j))$. If the cycle has no backtracking, then the corresponding endomorphism $\alpha \in \mathrm{End}(E(j))$ is primitive. Conversely, if $\alpha \in \mathrm{End}(E(j))$ is primitive and $\deg(\alpha) = \ell^e$ for some $e \in \mathbb{N}$, the cycle in $G(p, \ell)$ corresponding to α has no backtracking.*

Proof. The first statement is proved as Proposition 82 in [13]. His proof does not use the assumption that there are no repeated vertices in the cycle. Now assume that $\alpha \in \mathrm{End}(E(j))$ is primitive and $\deg(\alpha) = \ell^e$. Then by Proposition 10 of [10], the kernel of α is cyclic, generated by $P \in E[\ell^e]$. By Proposition 4.5, the cycle in $G(p, \ell)$ corresponding to α has no backtracking. $\qquad\square$

Suppose $\alpha \in \mathrm{End}(E(j))$ is an endomorphism of degree ℓ^e. We wish to describe what information we can infer about the order $\mathbb{Z}[\alpha]$ of $\mathbb{Q}(\alpha)$ from the cycle corresponding to α in $G(p, \ell)$. We will show that we can detect when $\mathbb{Z}[\alpha]$ is maximal at a prime above ℓ.

Lemma 4.7. *Let $\alpha \in \mathrm{End}(E(j))$ be a primitive endomorphism corresponding to a cycle $\{a_1, \ldots, a_e\}$ in $G(p, \ell)$ which begins at $E(j)$. Then a_1 is dual to a_e if and only if $\mathrm{tr}(\alpha) \equiv 0 \pmod{\ell}$.*

Proof. The endomorphism α determines an endomorphism $A = \alpha|_{E[\ell]}$ of $E[\ell]$. If $\mathrm{tr}(\alpha) \equiv 0 \pmod{\ell}$, then the characteristic polynomial of A is x^2. Thus $E[\ell] \subseteq \ker(\alpha^2)$, so α^2 is not primitive. Lemma 4.6 implies that the cycle

$\{a_1, \ldots, a_e, a_1, \ldots, a_e\}$ in $G(p, \ell)$ has backtracking, because the endomorphism corresponding to this cycle is α^2. We must have that a_1 is dual to a_e because α has no backtracking.

Conversely, assume a_1 is dual to a_e. Suppose that a_e is an edge from the vertex $E(j')$, and let $\phi_1 : E(j) \to E(j')$ be the isogeny corresponding to a_1 and let $\phi_e : E(j') \to E(j)$ be the isogeny corresponding to a_e. Then $\phi_e = \hat{\phi}_1 u$ for some $u \in \mathrm{Aut}(E(j))$. Thus $\alpha = \hat{\phi}_1 \alpha' \phi_1$, where α' is an endomorphism of $E(j')$. By Proposition 3.7, $\mathrm{tr}(\alpha) \equiv 0 \pmod{\ell}$. $\square$

This lets us conclude the following.

Lemma 4.8. *Let* $\{a_1, \ldots, a_e\}$ *be a cycle in* $G(p, \ell)$ *with no backtracking and such that* a_1 *is not dual to* a_e. *Suppose* a_1 *is an edge originating from* $E(j)$. *In the case that the cycle is a self-loop* a_1 *at* $E(j)$, *we assume that* a_1 *is not dual to itself. Let* $\alpha \in \mathrm{End}(E(j))$ *be the endomorphism corresponding to the cycle. Then the conductor of the quadratic order* $\mathbb{Z}[\alpha]$ *in* $\mathbb{Q}(\alpha)$ *is coprime to* ℓ.

Proof. As α is primitive, it determines a quadratic imaginary extension $\mathbb{Q}(\alpha)$ of $\mathbb{Q}$. The discriminant of α is $\mathrm{tr}(\alpha)^2 - 4\ell^e$, which is coprime to ℓ by Lemma 4.7. Thus the conductor of $\mathbb{Z}[\alpha]$, which divides the square part of the discriminant of α, is also coprime to ℓ. $\square$

Lemma 4.9. *Suppose that* $\alpha \in \mathrm{End}(E(j))$ *corresponds to a cycle* $\{a_1, \ldots, a_e\}$ *in* $G(p, \ell)$ *with no backtracking and such that* a_1 *is not dual to* a_e. *Let* $K = \mathbb{Q}(\alpha)$. *Then* ℓ *splits completely in* $\mathcal{O}_K$ *as* $\ell\mathcal{O}_K = \mathfrak{p}_1\mathfrak{p}_2$, *and* $\alpha\mathbb{Z}[\alpha] = (\mathfrak{p}_i \cap \mathbb{Z}[\alpha])^e$ *for* $i = 1$ *or* 2.

Proof. Since a_1 is not dual to a_e, the conductor of $\mathbb{Z}[\alpha]$ is coprime to ℓ by Lemmas 4.7 and 4.8. Let $\mathfrak{p}$ be a prime of $\mathcal{O}_K$ above ℓ. If ℓ ramifies in K, then the factorization $\alpha^2\mathcal{O}_K = \mathfrak{p}^{2e} = \ell^e\mathcal{O}_K$ implies that $\alpha^2\mathbb{Z}[\alpha] = \ell^e\mathbb{Z}[\alpha]$. But then $\alpha^2 = [\ell^e]\gamma$ for some $\gamma \in \mathbb{Z}[\alpha] \subseteq \mathrm{End}(E(j))$. On the other hand, α^2 must be primitive because the assumptions that α is primitive and a_e is not dual to a_1 imply that the cycle for α^2 has no backtracking. This implies that ℓ cannot ramify in K. If ℓ is inert, it follows that e is even and $\alpha\mathbb{Z}[\alpha] = \ell^{e/2}\mathbb{Z}[\alpha]$, again contradicting the assumption that α is primitive.

We conclude that ℓ must split completely in K, so let $\ell\mathcal{O}_K = \mathfrak{p}_1\mathfrak{p}_2$ be the factorization of $\ell\mathcal{O}_K$. We now claim that the ideal $\alpha\mathcal{O}_K$ factors as $\alpha\mathcal{O}_K = \mathfrak{p}_1^e$ or $\alpha\mathcal{O}_K = \mathfrak{p}_2^e$.

If the claim does not hold, then $\alpha\mathcal{O}_K = \mathfrak{p}_1^r\mathfrak{p}_2^s$ with $r, s > 0$. Without loss of generality we may assume that $r > s$. Then $\alpha\mathcal{O}_K = (\mathfrak{p}_1\mathfrak{p}_2)^s(\mathfrak{p}_1)^{r-s} = (\ell)^s(\mathfrak{p}_1)^{r-s}$. Then in $\mathbb{Z}[\alpha]$, we have the factorization

$$\alpha\mathbb{Z}[\alpha] = (\ell)^s(\mathfrak{p}_1 \cap \mathbb{Z}[\alpha])^{r-s}.$$

This implies that $\alpha = [\ell]\gamma$ for some $\gamma \in \mathrm{End}(E(j))$, but by Lemma 4.6, this contradicts the assumption that α has no backtracking. $\square$

Theorem 4.10. *Suppose that two cycles with no backtracking pass through $E(j)$ and that at least one cycle satisfies the hypotheses of Lemma 4.9. Denote the corresponding endomorphisms of $E(j)$ by α, β. Suppose further that α and β commute. Then there is a third cycle with no backtracking passing through $E(j)$ which corresponds to an endomorphism $\gamma \in \text{End}(E(j))$ and two automorphisms $u, v \in \text{Aut}(E(j))$ which commute with γ such that $\alpha = u\gamma^a$ and either $\beta = v\gamma^b$ or $\beta = v\hat{\gamma}^b$. In particular, the cycle for α is just the cycle for γ repeated a times, and the cycle for β is the cycle for γ or $\hat{\gamma}$ repeated b times.*

Proof. Assume that the cycle for α satisfies the assumption that its first edge is not dual to its last edge. Then the conductor of $\mathbb{Z}[\alpha]$ is coprime to ℓ. Since α and β commute, we must have $\beta \in \mathbb{Q}(\alpha)$. Let $\mathcal{O}$ be the order of $\mathbb{Q}(\alpha)$ whose conductor is the greatest common divisor of the conductors of $\mathbb{Z}[\alpha]$ and $\mathbb{Z}[\beta]$. Then $\mathcal{O} = \mathbb{Z}[\alpha] + \mathbb{Z}[\beta] \subseteq \text{End}(E(j))$ and the conductor of $\mathcal{O}$ is coprime to ℓ. By Lemma 4.9, ℓ splits completely in K; let $\mathfrak{p}_1, \mathfrak{p}_2$ be the primes above ℓ. Then without loss of generality, we have the factorization $\alpha\mathcal{O} = (\mathfrak{p}_1 \cap \mathcal{O})^i$ of $\alpha\mathcal{O}$ into primes of $\mathcal{O}$ by the same argument as in Lemma 4.9.

Observe that since $\beta \in \mathcal{O}$, $\hat{\beta} = \text{tr}(\beta) - \beta \in \mathcal{O}$. After possibly exchanging β with its dual $\hat{\beta}$, we get with the same argument that $\beta\mathcal{O} = (\mathfrak{p}_1 \cap \mathcal{O})^j$. Now let $d = \gcd(i, j)$, which implies that there exist $m, n \in \mathbb{Z}$ such that $d = im + jn$ and hence

$$(\mathfrak{p}_1 \cap \mathcal{O})^d = (\mathfrak{p}_1 \cap \mathcal{O})^{im+jn} = \alpha^m \beta^n \mathcal{O}.$$

Set $\gamma = \alpha^m \beta^n \in K$. Then

$$\gamma\mathcal{O} = (\mathfrak{p}_1 \cap \mathcal{O})^d$$

implies $\gamma \in \mathcal{O}$ and that γ must be primitive. Filtering the kernel of γ yields a cycle. Write $i = da$ and $j = db$ for $a, b \in \mathbb{N}$. Then $\gamma^a \mathcal{O} = \alpha\mathcal{O}$, so there exists $u \in \mathcal{O}^* \subseteq \text{Aut}(E(j))$ such that $\alpha = u\gamma^a$. We see that the cycle for α is just the cycle for γ repeated a times. Similarly, the cycle for β is the cycle for γ repeated b times. $\square$

We can state a more general result about when two cycles can give rise to commuting endomorphisms.

Corollary 4.11. *Suppose that $P = \{a_1, \ldots, a_m\}$ is a path in $G(p, \ell)$ without backtracking between $E(j)$ and $E(j')$ which does not pass through $E(0)$ or $E(1728)$. Suppose that $C = \{a_{m+1}, \ldots, a_{m+e}\}$ is a cycle beginning at $E(j')$ satisfying the assumptions of Lemma 4.9. Let $\hat{P}$ be a path $\{a_{m+e+1}, \ldots, a_{2m+e}\}$ without backtracking such that a_{m+1-k} is dual to a_{m+e+k} for $1 \leq k \leq m$. Let $\alpha \in \text{End}(E(j))$ be the endomorphism corresponding to the cycle $\{a_1, \ldots, a_{2m+e}\}$, the concatenation of P, C, and $\hat{P}$. Now let β be the endomorphism for another cycle in $G(p, \ell)$ without backtracking which starts at $E(j)$, and assume α and β commute. Then there exist automorphisms $u_1, u_2 \in \text{Aut}(E(j))$, an ℓ-power* *isogeny* $\phi : E(j) \rightarrow E(j')$, *an endomorphism* $\gamma \in \text{End}(E(j'))$, *automorphisms* $v_1, v_2 \in \text{Aut}(E(j'))$ *which commute with* γ, *and positive integers* a, b *such that* $\alpha = u_1 \circ \widehat{\phi} \circ v_1 \gamma^a \circ \phi$ *and* $\beta = u_2 \circ \widehat{\phi} \circ v_2 \gamma^b \circ \phi$ *or* $\beta = u_2 \circ \phi \circ v_2 \widehat{\gamma}^b \circ \phi$.

Proof. Let $\alpha' \in \text{End}(E(j'))$ be the endomorphism corresponding to C. Let the cycle for β be $\{a'_1, \ldots, a'_{e'}\}$. We can assume there is a positive integer n such that a'_k is dual to $a'_{e'-k+1}$ for $1 \leq k \leq n$, but a'_{n+1} is not dual to $a'_{e'-n}$. We can assume such an index exists because if not, a'_1 is not dual to $a'_{e'}$, and we could then apply the previous theorem to β. Write $f = e' - 2n$. Then we must have $f \geq 1$, because otherwise β will not be primitive. We then have two cases to consider: the cycle $\{a'_{n+1}, \ldots, a'_{n+f}\}$ satisfies the assumptions of Lemma 4.8, or $f = 1$ and a'_{n+1} is a self-loop which is dual to itself. We begin by considering the first case. We can assume that $m \leq n$, because otherwise we could swap the roles of α and β.

We will proceed by induction on m. Assume first that $m = 1$. Let β' correspond to $\{a'_2, \ldots, a'_{2n+f-1}\}$. If the cycle $\{a_1, \ldots, a_{e+2}, a'_1, \ldots, a'_{2n+f}\}$ has backtracking, it follows that a'_1 is dual to a_{e+2}. As the path $P = \{a_1\}$ does not pass through $E(0)$ or $E(1728)$, it follows that $\phi_{e+2} = \widehat{\psi_1}$ or $\phi_{e+2} = \widehat{\psi_1} \circ [-1]$. Additionally, since a_1 is dual to a_{e+2}, $\phi_{e+2} = \widehat{\phi_1}$ or $\phi_{e+2} = \widehat{\phi_1} \circ [-1]$. In any case, the equality $\alpha\beta = \beta\alpha$ implies that $\alpha'\beta' = \beta'\alpha'$. We can now apply Theorem 4.10. If $m > 1$, the corollary follows by applying the same argument to α' and β'.

We will now show that $\beta\alpha$ cannot be primitive. Assume that $\beta\alpha$ is primitive. Then its kernel is cyclic and thus contains a unique subgroup H of order ℓ with $H \subset E(j)[\ell]$. Then $H = \ker(\phi_1)$. On the other hand, the equality $\alpha\beta = \beta\alpha$ implies that $H = \ker(\psi_1)$, so we conclude $\phi_1 = \psi_1$ (here we use that ϕ_1 and ψ_1 are fixed representatives of edges in $G(p, \ell)$). This contradicts the assumption that $\beta\alpha$ is primitive, since a_1 is dual to a_e and $a'_1 = a_1$.

Now we consider the case that $f = 1$; we will show that in this case, β does not commute with α. Consider first the case that the cycle for β is just $\{a'_1\}$, a single self-loop which is dual to itself. Then $a_1 = a'_1$ by the same argument as above, by considering whether $\alpha\beta$ is primitive or not. Then the path $\{a_2, \ldots, a_{2m+e-1}\}$ is also a cycle beginning at $E(j)$, and its corresponding endomorphism also commutes with β. By induction we conclude then that β also commutes with $\{a_{m+1}, \ldots, a_{m+e}\}$, which is impossible by Theorem 4.10.

If now, in the cycle for β, we have $n < m$, we can use induction to conclude that $a_k = a'_k$ for $1 \leq k \leq n$, and then reduce to the case that β is a single self-loop dual to itself.

Thus we conclude that $m \leq n$. Again by using $\alpha\beta = \beta\alpha$, we find that $a_k = a'_k$ for $1 \leq k \leq m$, and we can reduce to the case of Theorem 4.10. $\qquad\square$

Corollary 4.12. *Suppose that two cycles* C_1 *and* C_2 *through* $E(j)$ *have no backtracking and that* C_1 *passes through a vertex through which* C_2 *does not pass. Suppose also that one cycle does not contain a self-loop which is dual to itself. Further assume that neither cycle passes through* $E(0)$ *or* $E(1728)$. *Then the corresponding endomorphisms in* $\text{End}(E(j))$ *are linearly independent.*

5 An Obstruction to Generating the Full Endomorphism Ring

If C is a cycle in $G(p, \ell)$ which passes through $E(j_1)$ and $E(j_2)$, then we can view it as starting at $E(j_1)$ or $E(j_2)$ and thus it corresponds to an endomorphism $\alpha \in \mathrm{End}(E(j_1))$ or $\alpha' \in \mathrm{End}(E(j_2))$. This suggests the following: suppose we have two cycles which have a path between $E(j_1)$ and $E(j_2)$ in common. Then we can view them as endomorphisms of each vertex. These endomorphisms generate an order $\mathcal{O}$ contained in the intersection of $\mathrm{End}(E(j_1))$ and $\frac{1}{\ell^e}\hat{\phi}\,\mathrm{End}(E(j_2))\phi$, where ϕ corresponds to the common path. These are two maximal orders inside of $\mathrm{End}(E(j_1)) \otimes \mathbb{Q}$ and thus the two cycles cannot generate a maximal order. However, this does not hold if $\mathrm{End}(E(j_1)) \simeq \mathrm{End}(E(j_2))$, i.e., j_1 is a Galois conjugate of j_2. This is formalized in the following theorem.

Theorem 5.1. *Suppose two cycles in $G(p, \ell)$ both contain the same path between two vertices $E(j_1)$ and $E(j_2)$. Let α and β be the corresponding endomorphisms of $E(j_1)$. If the path between $E(j_1)$ and $E(j_2)$ passes through additional vertices, or if $j_1^p \neq j_2$, then $\{1, \alpha, \beta, \alpha\beta\}$ is not a basis for $\mathrm{End}(E(j_1))$.*

Proof. We can assume that $j_1^p \neq j_2$, by replacing j_2 with an earlier vertex in the path if necessary. Let the path from $E(j_1)$ to $E(j_2)$ be correspond to the isogeny $\phi : E(j_1) \to E(j_2)$. By assumption, we can write $\alpha = \alpha_1\phi$ and write $\beta = \beta_1\phi$. Let $\alpha' = \phi\alpha_1$ and $\beta' = \phi\beta_1$ be the corresponding endomorphisms of $E(j_2)$. Assume towards contradiction that $\langle 1, \alpha, \beta, \alpha\beta \rangle = \mathrm{End}(E(j_1))$. Denote the lists

$$\{x_0, x_1, x_2, x_3\} = \{1, \alpha, \beta, \alpha\beta\}$$
$$\{y_0, y_1, y_2, y_3\} = \{1, \alpha', \beta', \alpha'\beta'\}.$$

We now show that $\mathrm{tr}(x_i\widehat{x_j}) = \mathrm{tr}(y_i\widehat{y_j})$ for $i, j = 0, \ldots, 3$. Observe that $[\deg\phi](x_i\widehat{x_j}) = \widehat{\phi}y_i\widehat{y_j}\phi$, so

$$\deg(\phi)\,\mathrm{tr}(x_i\widehat{x_j}) = \mathrm{tr}(\widehat{\phi}y_i\widehat{y_j}\phi).$$

On the other hand, we use Lemma 3.7 to compute

$$\mathrm{tr}(\widehat{\phi}y_i\widehat{y_j}\phi) = \deg(\phi)\,\mathrm{tr}(y_i\widehat{y_j}).$$

This implies that the embedding

$$\mathrm{End}(E(j_2)) \hookrightarrow \mathrm{End}(E(j_1)) \otimes \mathbb{Q}$$
$$\rho \mapsto \widehat{\phi}\rho\phi \otimes \frac{1}{\deg\phi}$$

maps $\langle 1, \alpha', \beta', \alpha'\beta' \rangle$ to an order isomorphic to $\mathrm{End}(E(j_1))$ by [17, Corollary 4.4]. But this violates Deuring's correspondence.   □

One might conjecture that two cycles in $G(p, \ell)$ which only intersect at one vertex $E(j)$ generate $\mathrm{End}(E(j))$, but the example in the following section shows this might not be true. In particular, there is an example of two cycles which generate an order $\mathcal{O}$ which is not maximal, but there is a unique maximal order containing $\mathcal{O}$.

6   Examples

We used the software package Magma to perform most of the computations required to compute the endomorphism rings of supersingular elliptic curves in characteristic p with $p \in \{31, 101, 103\}$. In all cases we worked with the 2-isogeny graph. We started with the supersingular j-invariants and found models for the elliptic curves $E(j)$ as in Equation 2.1.1 that we transformed into ones of the form $y^2 = x^3 + ax + b$ for some $A, B \in \mathbb{F}_{p^2}$. Then for every $E(j)$ we computed the 2-torsion points to generate its 2-isogenies, as in Section 2.1.

By Theorem 3.2 and Lemma 3.3 we know that $\mathrm{End}(E(j))$ corresponds to a maximal order in $B_{p,\infty}$.

For each vertex corresponding to $E(j)$, we select cycles in the 2-isogeny graph that satisfy the conditions of Theorem 4.10 and compute their traces and norms. Then we find elements of $B_{p,\infty}$ with these traces and norms and verify that they generate a maximal order.

Example 6.1 ($p = 31$). Let $p = 31$. The unique quaternion algebra ramified at p and ∞ is

$$B_{31,\infty} = \mathbb{Q} + \mathbb{Q}i + \mathbb{Q}j + \mathbb{Q}ij,$$

where $i^2 = -1$ and $j^2 = -31$.

There are three j-invariants corresponding to isomorphism classes of supersingular elliptic curves over $\mathbb{F}_{p^2}$, namely 2, 4, and 23. Figure 1 shows the 2-isogeny graph with labeled edges.

Table 1 contains, for each vertex, two cycles that correspond to elements that generate a maximal order in $B_{p,\infty}$. Hence these two cycles must generate the full endomorphism ring.

Fig. 1 2-isogeny graph for $p = 31$.

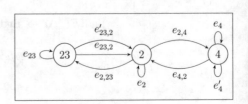

Table 1 Pairs of cycles that generate maximal orders in $B_{31,\infty}$

Vertex	Cycle	Trace	Norm
2	e_2	0	2
	$e_{2,4}e_4e_{4,2}$	2	8
4	e_4	1	2
	$e_{4,2}e_2e_{2,4}$	0	8
23	e_{23}	2	2
	$e_{23,2}e_2e_{2,23}$	−1	8

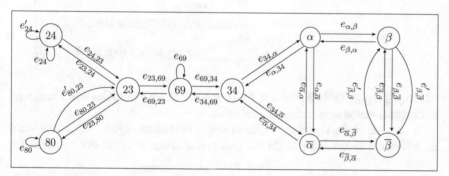

Fig. 2 2-isogeny graph for $p = 103$.

With this data we are able to generate the maximal orders that correspond to each endomorphism ring:

$$\mathrm{End}(E(23)) \cong \left\langle 1, -i, -\frac{1}{2}i + \frac{1}{2}ij, \frac{1}{2} - \frac{1}{2}j \right\rangle,$$

$$\mathrm{End}(E(2)) \cong \left\langle 1, \frac{1}{4}i\frac{1}{4}ij, 2i, \frac{1}{2} - \frac{1}{2}j \right\rangle,$$

$$\mathrm{End}(E(4)) \cong \left\langle 1, \frac{1}{2} + \frac{1}{6}i + \frac{1}{6}j - \frac{1}{6}ij, \frac{5}{6}i + \frac{1}{3}j + \frac{1}{6}ij, -\frac{13}{6}i + \frac{1}{3}j + \frac{1}{6}ij \right\rangle.$$

Example 6.2 ($p = 103$). Let $p = 103$. The unique quaternion algebra ramified at p and ∞ is

$$B_{103,\infty} = \mathbb{Q} + \mathbb{Q}i + \mathbb{Q}j + \mathbb{Q}ij,$$

where $i^2 = -1$ and $j^2 = -103$.

The supersingular j-invariants over $\mathbb{F}_{p^2}$ are 23, 24, 69, 34, 80, and four defined over $\mathbb{F}_{p^2} - \mathbb{F}_p$: α, β, and their conjugates. Figure 2 shows the 2-isogeny graph.

After several computations, we were able to find generators for the maximal orders corresponding to all the endomorphism rings of supersingular curves $E(j)$,

Table 2 Pairs of cycles that generate maximal orders in $B_{103,\infty}$

Vertex	Cycle	Trace	Norm
34	$e_{34,\overline{\alpha}}e_{\overline{\alpha},\alpha}e_{\alpha,34}$	-3	8
	$e_{34,69}e_{69}e_{69,34}$	0	8
69	e_{69}	0	2
	$e_{69,34}e_{34,\alpha}e_{\alpha,\overline{\alpha}}e_{\overline{\alpha},34}e_{34,69}$	-6	32
23	$e_{23,24}e_{24}e_{24,23}$	2	8
	$e_{23,80}e_{80}e_{80,23}$	-4	8
80	e_{80}	2	2
	$e_{80,23}e_{23,69}e_{69}e_{69,23}e_{23,80}$	0	32
24	e_{24}	-1	2
	$e_{24,23}e_{23,69}e_{69}e_{69,23}e_{23,24}$	0	32

where $j \in \mathbb{F}_{103^2}$. Table 2 contains, for each such vertex two cycles that correspond to elements that generate the maximal order.

In the case of the vertex α, we found an example of two cycles that do not share an additional vertex but that do not generate a maximal order. For instance, the cycles

$$e_{\alpha,\beta}e'_{\beta,\overline{\beta}}\overline{e}'_{\overline{\beta},\beta}e_{\beta,\alpha}$$

$$e_{\alpha,34}e_{34,69}e_{69}e_{69,34}e_{34,\alpha}$$

generate the order $\mathcal{O} = \left\langle 1, -\frac{1}{2}+\frac{17}{6}i-\frac{1}{6}j+\frac{1}{6}ij, -\frac{5}{2}i+\frac{1}{2}ij, -\frac{1}{2}-\frac{22}{3}i-\frac{11}{6}j-\frac{2}{3}ij \right\rangle$ and there is a unique maximal order containing it, hence this corresponds to $\mathrm{End}(E(\alpha)) \cong \mathrm{End}(E(\overline{\alpha}))$. Finally, there is only one maximal order remaining in $B_{103,\infty}$, which by Theorem 3.2 is isomorphic to the endomorphism rings of $E(\beta)$ and $E(\overline{\beta})$.

The endomorphism rings are then isomorphic to the following maximal orders:

$$\mathrm{End}(E(80)) \cong \left\langle 1, i, \frac{1}{2}i + \frac{1}{2}ij, \frac{1}{2} + \frac{1}{2}j \right\rangle,$$

$$\mathrm{End}(E(23)) \cong \left\langle 1, 2i, \frac{3}{4}i + \frac{1}{4}ij, \frac{1}{2} - \frac{1}{2j} \right\rangle,$$

$$\mathrm{End}(E(34)) \cong \left\langle 1, \frac{17}{14}i + \frac{1}{14}ij\,\frac{15}{7}i - \frac{2}{7}ij, \frac{1}{2} - \frac{1}{2}j \right\rangle,$$

$$\mathrm{End}(E(69)) \cong \left\langle 1, \frac{1}{2} + \frac{1}{7}i + \frac{3}{14}j, \frac{1}{2} - \frac{16}{7}i + \frac{1}{14}j, \frac{1}{2} - \frac{17}{14}i - \frac{1}{14}j - \frac{1}{2}ij \right\rangle,$$

$$\mathrm{End}(E(24)) \cong \left\langle 1, \frac{1}{2} + \frac{3}{8}i + \frac{1}{8}ij, \frac{1}{2} - \frac{29}{8}i + \frac{1}{8}ij, -\frac{13}{8}i + \frac{1}{2}j + \frac{1}{8}ij \right\rangle,$$

$$\mathrm{End}(E(\alpha)) \cong \left\langle -1, -\frac{1}{2}+\frac{1}{6}i-\frac{1}{6}j-\frac{1}{6}ij, 3i, \frac{5}{6}i-\frac{1}{3}j+\frac{1}{6}ij \right\rangle,$$

$$\mathrm{End}(E(\beta)) \cong \left\langle 1, \frac{1}{2}+\frac{13}{10}i+\frac{1}{10}j-\frac{1}{10}ij, -\frac{12}{5}i+\frac{1}{5}j-\frac{1}{5}ij, \frac{1}{2}-\frac{3}{5}i+\frac{3}{10}j+\frac{1}{5}ij \right\rangle.$$

Example 6.3 ($p = 101$). Let $p = 101$. The unique quaternion algebra ramified at p and ∞ is

$$B_{101,\infty} = \mathbb{Q} + \mathbb{Q}i + \mathbb{Q}j + \mathbb{Q}ij,$$

where $i^2 = -2$ and $j^2 = -101$.

The supersingular j-invariants over $\mathbb{F}_{101^2}$ are $64, 0, 21, 57, 3, 59, 66$, and two additional ones, which we denote by α and $\overline{\alpha}$, are defined over $\mathbb{F}_{p^2} - \mathbb{F}_p$. Figure 3 shows the 2-isogeny graph.

It was possible to find two cycles that generate the maximal order corresponding to $\mathrm{End}(E(j))$, where $j \in \{3, 59, 64, 66\}$. Table 3 contains the data for these cycles. For the vertices $21, 57, \alpha$ no two cycles were found that generate the full endomorphism ring. However, in each of these cases we were able to generate an order from two cycles which happened to be contained in a unique maximal order. These cycles are listed in Table 4.

By Theorem 5.1, no two cycles through $j = 0$ generate a maximal order, but it is possible to determine which one corresponds to the endomorphism ring of $E(0)$ once we ruled out the other seven. The endomorphism rings are then isomorphic to the following maximal orders:

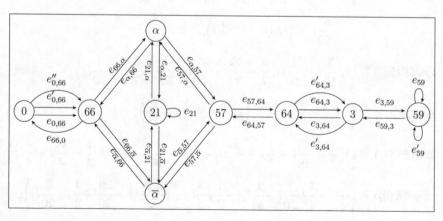

Fig. 3 2-isogeny graph for $p = 101$.

Table 3 Pairs of cycles that generate maximal orders in $B_{101,\infty}$

Vertex	Cycle	Trace	Norm
3	$e_{3,59}e_{59}e_{59,3}$	2	8
	$e_{3,64}e'_{64,3}$	-1	4
59	e_{59}	-1	2
	$e_{59,3}e_{3,64}e_{64,3}e_{3,59}$	-8	16
64	$e_{64,57}e_{57,\alpha}e_{\alpha,66}e_{66,\overline{\alpha}}e_{\overline{\alpha},57}e_{57,64}$	10	64
	$e_{64,3}e'_{3,64}$	-1	4
66	$e_{66,0}e_{0,66}$	2	4
	$e_{66,\alpha}e_{\alpha,57}e_{57,\overline{\alpha}}e_{\overline{\alpha},66}$	5	16

Table 4 Pairs of cycles that generate non-maximal orders in $B_{101,\infty}$

Vertex	Cycle	Trace	Norm
21	e_{21}	0	2
	$e_{21,\alpha}e_{\alpha,66}e_{66,0}e'_{0,66}e_{66,\alpha}e_{\alpha,21}$	-8	64
57	$e_{57,64}e_{64,3}e_{3,59}e_{59}e_{59,3}e_{3,64}e_{64,57}$	-8	128
	$e_{57,\alpha}e_{\alpha,66}e_{66,\overline{\alpha}}e_{\overline{\alpha},37}$	-5	16
α	$e_{\alpha,21}e_{21}e_{21,\overline{\alpha}}e_{\overline{\alpha},57}e_{57,\alpha}$	5	32
	$e_{\alpha,66}e_{66,0}e'_{0,66}e_{66,\alpha}$	4	16

$$\mathrm{End}(E(3)) \cong \left\langle 1, \frac{1}{2} - \frac{13}{12}i + \frac{1}{12}ij, \frac{5}{6}i + \frac{1}{6}ij, \frac{5}{12}i - \frac{1}{2}j + \frac{1}{12}ij \right\rangle,$$

$$\mathrm{End}(E(59)) \cong \left\langle 1, \frac{1}{2} + \frac{5}{12}i - \frac{1}{12}ij, -\frac{13}{6}i - \frac{1}{6}ij, -\frac{13}{12}i + \frac{1}{2}j - \frac{1}{12}ij \right\rangle,$$

$$\mathrm{End}(E(64)) \cong \left\langle -1, -\frac{1}{2} - \frac{3}{5}i - \frac{1}{10}j + \frac{1}{10}ij, -\frac{1}{2} - \frac{21}{20}i + \frac{1}{5}j + \frac{1}{20}ij, \right.$$
$$\left. -\frac{67}{20}i - 1/10j - 3/20ij \right\rangle,$$

$$\mathrm{End}(E(66)) \cong \left\langle 1, \frac{7}{10}i - \frac{1}{10}ij, \frac{1}{2} - \frac{29}{20}i - \frac{3}{20}ij, \frac{7}{20}i - \frac{1}{2}j - \frac{1}{20}ij \right\rangle,$$

$$\mathrm{End}(E(21)) \cong \left\langle -1, i, -\frac{1}{2} + \frac{1}{4}i - \frac{1}{4}ij, -\frac{1}{2} + \frac{1}{2}i - \frac{1}{2}j \right\rangle,$$

$$\mathrm{End}(E(57)) \cong \left\langle 1, \frac{1}{2} - \frac{13}{28}i + \frac{1}{7}j + \frac{1}{28}ij, -\frac{53}{28}i - \frac{1}{14}j + \frac{3}{28}ij, \frac{1}{2} - \frac{11}{4}i - \frac{1}{4}ij \right\rangle,$$

$$\mathrm{End}(E(0)) \cong \left\langle -1, -\frac{1}{2} + \frac{7}{20}i + \frac{1}{20}ij, -\frac{1}{2} + \frac{9}{5}i + \frac{1}{2}j - \frac{1}{10}ij, -\frac{29}{20}i + \frac{1}{2}j + \frac{3}{20}ij \right\rangle,$$

$$\mathrm{End}(E(\alpha)) \cong \mathrm{End}(E(\overline{\alpha})) \cong \left\langle -1, 2i, -\frac{1}{2} + \frac{3}{8}i + \frac{1}{4}j - \frac{1}{8}ij, -\frac{7}{8}i + \frac{1}{4}j + \frac{1}{8}ij \right\rangle.$$

Appendix A: Modified Schoof's Algorithm for Traces of Arbitrary Endomorphisms

Let E be an elliptic curve over a finite field $\mathbb{F}_q$ of characteristic $p \neq 2, 3$. The Frobenius endomorphism $\phi \in \text{End}_{\mathbb{F}_q}(E)$ takes any point $(x, y) \in E(\mathbb{F}_q)$ to (x^q, y^q); it satisfies the relation in $\text{End}_{\mathbb{F}_q}(E)$, given by

$$\phi^2 - t\phi + q = 0.$$

Here, t is called the trace of the Frobenius endomorphism, and it is related to the number of $\mathbb{F}_q$-points on E via the relation

$$\#E(\mathbb{F}_q) = q + 1 - t.$$

Schoof's algorithm [20] computes the trace of the Frobenius endomorphism in $O(\log^9 q)$ elementary operations (bit operations). This algorithm has been improved in [23] to be completed in $O(\log^5 q \log \log q)$ operations.

Let E be a supersingular elliptic curve defined over $\mathbb{F}_{p^2}$. Here we outline a modification of Schoof's algorithm that computes the trace of any endomorphism $\alpha \in \text{End}_{\mathbb{F}_q}(E)$ that corresponds to a cycle in the ℓ-isogeny graph, where $\ell \neq p$ is a prime. That is, we assume that we are given a cycle of length e in the ℓ-isogeny graph; this path can be represented as a chain of e isogenies of degree ℓ, $\phi_k : E_k \to E_{k+1}$ for $k = 0, \ldots, e - 1$. Here $E_0, \ldots, E_e$ are elliptic curves in short Weierstrass form, defined over $\mathbb{F}_{p^2}$, and $E_0 = E_e$. We assume the isogenies are specified by their rational maps. We remark that if this cycle is instead represented by a sequence of ℓ-isogenous elliptic curves, then one can compute a corresponding sequence of ℓ-isogenies in $\tilde{O}(n^2)$ time by Theorem 2 of [3], where $n = \max\{\lceil \log p \rceil, \ell, e\}$. In the context we are interested in (where p is of cryptographic size, $\ell = O(\log p)$, and we assume $e = O(\log p)$), we observe that finding a cycle in $G(p, \ell)$ could require time exponential in $\log p$, so we may as well assume that we are given the isogenies.

More precisely then, we assume that the input to our algorithm is a cycle of isogenies, each given explicitly as in Proposition 4.1 of [3] which we record here.

Proposition A.1. *Let $E : y^2 = x^3 + Ax + B$ be an elliptic curve. Then every (normalized) ℓ-isogeny $\psi : E \to E'$ can be written as*

$$\psi(x, y) = \left(\frac{N(x)}{D(x)}, y \left(\frac{N(x)}{D(x)} \right)' \right),$$

where

$$D(x) = \prod_{P \in \ker \psi \setminus \{0\}} (x - x_P)$$

and we define $N(x)$ by the relation

$$\frac{N(x)}{D(x)} = \ell x - \sigma - (3x^2 + A)\frac{D'(x)}{D(x)} - 2(x^3 + Ax + B)\left(\frac{D'(x)}{D(x)}\right)'.$$

Here, σ is the coefficient of $x^{\ell-1}$ in $D(x)$, the sum of the abscissas of the nonzero points of the kernel of ψ.

Proof. This is Proposition 4.1 of [3]. □

By Corollary 2.5, if E is defined over $\mathbb{F}_{p^2}$ we can take these isogenies to be defined over an extension of degree at most degree 6 of $\mathbb{F}_{p^2}$. If $\ell = O(\log p)$ and the path has length $e = O(\log p)$, which are the parameters that are most interesting, we will show that the trace of this endomorphism can be computed in $\tilde{O}(\log^7 p)$ time by using a modified version of Schoof's algorithm, where we use $f(n) = \tilde{O}(g(n))$ to mean that there exists k such that $f(n) = O(g(n)\log^k n)$.

The naïve computation of the composition of the e isogenies via Vélu's formula yields a formula for the ℓ^e-isogeny that requires at least $O(\ell^e)$ elementary operations; in order to cut down on the number of elementary operations required to compute the explicit formula for the isogeny, we note that the explicit isogeny formula is simpler on the set of m-torsion points for any m, by taking the quotient modulo the division polynomials. Thus, ℓ^e-isogenies on $E[m]$ can be computed much more quickly, and this is sufficient information to which one can apply Schoof's idea. We remark that the algorithm will correctly compute the trace of an endomorphism of an ordinary curve $E/\mathbb{F}_q$, but unlike in the supersingular case and without further assumptions on the cycle, not all of the isogenies are defined over $\mathbb{F}_q$ (or an extension of $\mathbb{F}_q$ of bounded degree).

A.1. Complexity of Computing Endomorphisms on m-Torsion

Let $f_k(X)$ denote the k-th division polynomial of E. It is the polynomial whose roots are the x-coordinates of the nonzero elements of the k-torsion subgroup of E. When k is coprime to p, the degree of f_k is $(k^2 - 1)/2$. The division polynomials can be defined recursively and the complexity of computing them is analyzed in [23].

Let $M(n)$ denote the number of elementary operations required to multiply two n-bit integers. If we choose to multiply two n-bit integers via long multiplication, then $M(n) = O(n^2)$; if we multiply two numbers using the Fast Fourier Transform (FFT), then $M(n) = O(n\log n \log\log n)$.

Proposition A.2. *Given a natural number $m > 1$, the division polynomials $f_1, \ldots, f_m$ can be computed in $O(mM(m^2 \log q))$ time.*

Proof. Using the recursive relations defining the division polynomials, f_k can be computed in $O(M(k^2 \log q))$ time by using a double-and-add method. Thus $f_1, \ldots, f_m$ can be computed in $O(mM(m^2 \log q))$ time; see [23, Section 5.1]. □

We continue to work over $\mathbb{F}_q$; typically we will work over an extension of $\mathbb{F}_{p^2}$ of degree at most 6.

Given an ℓ-isogeny $\psi : E \to E'$ as well as a prime $m \neq 2, p$, we are interested in the explicit formula for the induced isogeny on the m-torsion points $\psi_m : E[m] \to E'[m]$. If E is defined by the equation $y^2 = x^3 + ax + b$, and $f_m(x)$ is the m-th division polynomial for E, then $E[m] = \operatorname{Spec} \mathbb{F}_q[x, y]/I$, where $I = \langle f_m(x), y^2 - (x^3 + ax + b) \rangle$. Thus we may reduce the coordinates of the explicit formula for the isogeny ψ given by $(x, y) \mapsto (X(x, y), Y(x, y))$ modulo the ideal I, and the resulting map ψ_m agrees with ψ on $E[m]$. Let $d = \max m, \ell$.

Proposition A.3. *Keeping the notation of the discussion in the above paragraph, $\deg \psi_m = O(d)$, and ψ_m can be computed in $O(M(d^2 \log q) \log d)$ elementary operations.*

Proof. First we observe that by Proposition A.1, the rational functions which define ψ have degree $O(\ell)$. Next, reduce modulo $f_m(x)$, so that the degree of the resulting expression is bounded by $\deg f_m = O(m^2)$. Then by [23, Lemma 9, p. 315], it takes $O(M(d^2 \log q) \log d)$ elementary operations to compute the reduction of the isogeny formula modulo f_m.   □

A.2. Computing the Trace on m-Torsion Points

To compute the trace of an endomorphism $\psi \in \operatorname{End}(E)$, where ψ appears as a cycle of length e in the supersingular ℓ-isogeny graph in characteristic p, we will compute $\operatorname{tr}(\psi) \pmod{m}$ for several primes m and then recover the trace using the Chinese Remainder Theorem, as in Schoof's algorithm.

The endomorphism ψ satisfies the equation $x^2 - \operatorname{tr}(\psi)x + \operatorname{norm}(\psi)$. There is a simple relationship between $\operatorname{tr}(\psi)$ and $\operatorname{norm}(\psi)$:

Lemma A.4. *Let $\psi \in \operatorname{End}(E)$. Then $|\operatorname{tr}(\psi)| \leq 2 \operatorname{norm}(\psi)$.*

Proof. If ψ is multiplication by some integer, then its characteristic polynomial is $x^2 \pm 2nx + n^2$, with $n \in \mathbb{N}$. Then $|\operatorname{tr}(\psi)| = 2n$, $\operatorname{norm}(\psi) = n^2$, and the statement of the lemma holds.

If ψ is not multiplication by an integer, then $\mathbb{Z}[\psi]$ is an order in the ring of integers $\mathcal{O}_K$ for some quadratic imaginary number field K. Hence we can fix an embedding $\iota : \mathbb{Z}[\psi] \hookrightarrow \mathcal{O}_K$. Since $\iota(\psi)$ is imaginary, its characteristic polynomial $x^2 - \operatorname{tr}(\psi)x + \operatorname{norm}(\psi)$ must have discriminant < 0, so $|\operatorname{tr}(\psi)| \leq 2\sqrt{\operatorname{norm}(\psi)}$.   □

As in Schoof's algorithm, we begin by looking for a bound L such that

$$N := \prod_{\substack{m \leq L \text{ prime} \\ m \neq 2, p}} m > 2 \operatorname{norm}(\psi) = 2\ell^e, \tag{A.2.1}$$

where the last equality follows from the fact that the cycle corresponding to ψ in the isogeny graph has length e, so $\text{norm}(\psi) = \ell^e$. By the prime number theorem, we can take $L = O(\log p)$ and there are $O(\log p / \log \log p)$ many primes less than L.

Let m be a prime. Any $\psi \in \text{End}(E)$ induces an endomorphism $\psi_m \in \text{End}(E[m])$; if ψ_m has characteristic polynomial $x^2 - t_m x + n_m$, then $t_m \equiv \text{tr}(\psi)$ (mod m). After computing t (mod m) for each $m < L$, we can compute t (mod N) using the Chinese Remainder Theorem. The bound in Lemma A.4 then lets us compute the value of $\text{tr}(\psi)$. Now, fix one such prime m.

A.2.1. Computation of $\text{tr}(\psi_m)$

Let $t_m \equiv \text{tr}(\psi) \bmod m$. Then the relation $\psi_m^2 - t_m \psi_m + n_m = 0$ holds in $\text{End}(E[m]) := \text{End}(E)/(m)$. Here, $n_m \equiv \text{norm}(\psi_m) = \ell^e \bmod m$, with $0 \le n_m < m$.

Furthermore, one has an explicit formula for $\psi_m : E[m] \to E[m]$ by reducing the explicit coordinates for ψ modulo the ideal I (using the notation in the discussion before Proposition A.3), with $\deg \psi_m = O(m^2)$. Using the addition formulas for E, we can compute the explicit formula for $\psi_m^2 + n_m$, and reduce it to modulo I. The main modification to Schoof's algorithm, as it is described in [23, 5.1], is to replace the Frobenius endomorphism on $E[m]$ with ψ_m. Having computed $\psi_m^2 + n_m$ and ψ_m, for τ with $0 \le \tau \le m - 1$ we compute $\tau \psi_m$ until

$$\psi_m^2 + n_m = \tau \psi_m$$

in $\text{End}(E[m])$. Then $\tau = t_m$. Having computed t_m for sufficiently many primes, we recover $\text{tr}\,\psi$ using the Chinese Remainder Theorem.

A.2.2. Complexity Analysis for Computing the Trace

Proposition A.5. *Let $E/\mathbb{F}_q$ be a supersingular elliptic curve. Let ψ be an isogeny of E of degree ℓ^e, specified as a chain $\phi_1, \ldots, \phi_e$ of ℓ-isogenies, whose explicit formulas are given. The explicit formula for ψ_m can be computed in $O(edM(d \log q) \log d)$ time, where $d \in \max\{\ell, m^2\}$.*

Proof. The expression for ψ_m can be computed by computing $(\phi_k)_m$ for $k = 1, \ldots, e$, composing the rational maps, and reducing modulo I at each step. The calculation of $f \circ g \bmod h$, where $f, g, h \in \mathbb{F}_q[x]$ are polynomials of degree at most d, takes $O(dM(d \log q))$ elementary operations using the naïve approach. Thus, computing e of these compositions, reducing modulo f_m at each step, takes $O(edM(d \log q) \log q)$ time.   □

We now wish to compute the trace of an endomorphism of E corresponding to a cycle in $G(p, \ell)$. Since the diameter of $G(p, \ell)$ is $O(\log p)$, we are interested in computing the trace of a cycle of length $e = O(\log p)$ in $G(p, \ell)$. We are also interested in the case where ℓ is a small prime, so we will take $\ell = O(\log p)$. The resulting generalization of Schoof's algorithm runs in time polynomial in $\log p$.

Theorem A.6. *Let* $p > 3$ *be a prime and let* ψ *be an endomorphism of a supersingular elliptic curve* $E/\mathbb{F}_{p^2}$ *given as a chain of* ℓ-*isogenies,*

$$\psi = \phi_e \circ \cdots \circ \phi_1,$$

where each ϕ_k *is specified by its rational functions and is defined over* $\mathbb{F}_q$. *We can take* $\mathbb{F}_q$ *to be an extension of* $\mathbb{F}_{p^2}$ *of degree at most 6. Let* $n = \lceil \log p \rceil$ *and assume* $e, \ell = O(n)$. *Then the modified version of Schoof's algorithm computes* $\mathrm{tr}\,\psi$ *in* $\tilde{O}(n^7)$ *time.*

Proof. We follow the steps in our modification of Schoof's algorithm. Since norm $\psi = \ell^e$, we first choose a bound $L = O(\log \ell^e)$.

We can compute ψ_m in time $\tilde{O}(n^6)$ time by Proposition A.5. For a prime $m < L$, we compute $\mathrm{tr}\,\psi_m$, the trace of the induced isogeny ψ_m on $E[m]$, by reducing by the m-division polynomial f_m whenever possible.

Having computed ψ_m and ψ_m^2, with the same argument as in the proof of Theorem 10 of [23], we can compute t_m in $O((m + \log q)(M(m^2 \log q)))$ time. This is because once ψ_m and ψ_m^2 are computed, the algorithm proceeds the same way as Schoof's original algorithm. We must repeat this $L = O(\log p) = O(n)$ times.

Once we compute $\mathrm{tr}\,\psi_m$ for each prime $m \neq p$ less than L, we compute $\mathrm{tr}\,\psi$ using the Chinese Remainder Theorem. This step is dominated by the previous computations. Thus we have a total run time of $\tilde{O}(n^7)$. $\qquad\square$

References

1. Reza Azarderakhsh, Matthew Campagna, Craig Costello, Luca De Feo, Basil Hess, Amir Jalali, David Jao, Brian Koziel, Brian LaMacchia, Patrick Longa, Michael Naehrig, Joost Renes, Vladimir Soukharev, and David Urbanik. Supersingular isogeny key encapsulation. Submission to the NIST Post-Quantum Standardization project, 2017. https://csrc.nist.gov/Projects/Post-Quantum-Cryptography/Round-1-Submissions.
2. Jean-François Biasse, David Jao, and Anirudh Sankar. A quantum algorithm for computing isogenies between supersingular elliptic curves. In *Progress in cryptology—INDOCRYPT 2014*, volume 8885 of *Lecture Notes in Comput. Sci.*, pages 428–442. Springer, Cham, 2014.
3. A. Bostan, F. Morain, B. Salvy, and É. Schost. Fast algorithms for computing isogenies between elliptic curves. *Math. Comp.*, 77(263):1755–1778, 2008.
4. J. M. Cerviño. Supersingular elliptic curves and maximal quaternionic orders. In *Mathematisches Institut, Georg-August-Universität Göttingen: Seminars Summer Term 2004*, pages 53–60. Universitätsdrucke Göttingen, Göttingen, 2004.
5. Ilya Chevyrev and Steven D. Galbraith. Constructing supersingular elliptic curves with a given endomorphism ring. *LMS J. Comput. Math.*, 17(suppl. A):71–91, 2014.
6. Denis X. Charles, Eyal Z. Goren, and Kristin Lauter. Cryptographic hash functions from expander graphs. *J. Cryptology*, 22(1):93–113, 2009.

7. Max Deuring. Die Typen der Multiplikatorenringe elliptischer Funktionenkörper. *Abh. Math. Sem. Hansischen Univ.*, 14:197–272, 1941.
8. Luca De Feo, David Jao, and Jérôme Plût. Towards quantum-resistant cryptosystems from supersingular elliptic curve isogenies. *J. Math. Cryptol.*, 8(3):209–247, 2014.
9. Christina Delfs and Steven D. Galbraith. Computing isogenies between supersingular elliptic curves over $\mathbb{F}_p$. *Des. Codes Cryptogr.*, 78(2):425–440, 2016.
10. Kirsten Eisenträger, Sean Hallgren, Kristin Lauter, Travis Morrison, and Christophe Petit. Supersingular isogeny graphs and endomorphism rings: reductions and solutions. *Eurocrypt 2018, LNCS 10822*, pages 329–368, 2018.
11. Steven D. Galbraith, Christophe Petit, and Javier Silva. Identification protocols and signature schemes based on supersingular isogeny problems. In Tsuyoshi Takagi and Thomas Peyrin, editors, *Advances in Cryptology – ASIACRYPT 2017*, pages 3–33, Cham, 2017. Springer International Publishing.
12. David Kohel, Kristin Lauter, Christophe Petit, and Jean-Pierre Tignol. On the quaternion l-isogeny path problem. *LMS Journal of Computation and Mathematics*, 17:418–432, 2014.
13. David Kohel. *Endomorphism rings of elliptic curves over finite fields*. PhD thesis, University of California, Berkeley, 1996.
14. Kristin Lauter and Ken McMurdy. Explicit generators of endomorphism rings of supersingular elliptic curves. Preprint, 2004.
15. Ken McMurdy. Explicit representation of the endomorphism rings of supersingular elliptic curves. https://phobos.ramapo.edu/~kmcmurdy/research/McMurdy-ssEndoRings.pdf, 2014.
16. J.-F. Mestre. La méthode des graphes. Exemples et applications. In *Proceedings of the international conference on class numbers and fundamental units of algebraic number fields (Katata, 1986)*, pages 217–242. Nagoya Univ., Nagoya, 1986.
17. Gabriele Nebe. Finite quaternionic matrix groups. *Represent. Theory*, 2:106–223, 1998.
18. NIST. Post-quantum cryptography, 2016. csrc.nist.gov/Projects/Post-Quantum-Cryptography; accessed 30-September-2017.
19. Arnold Pizer. An algorithm for computing modular forms on $\Gamma_0(N)$. *J. Algebra*, 64(2):340–390, 1980.
20. René Schoof. Elliptic curves over finite fields and the computation of square roots mod p. *Math. Comp.*, 44(170):483–494, 1985.
21. René Schoof. Counting points on elliptic curves over finite fields. *J. Théor. Nombres Bordeaux*, 7(1):219–254, 1995. Les Dix-huitièmes Journées Arithmétiques (Bordeaux, 1993).
22. J.H. Silverman. *The Arithmetic of Elliptic Curves*. Graduate Texts in Mathematics. Springer New York, 2009.
23. Igor E. Shparlinski and Andrew V. Sutherland. On the distribution of Atkin and Elkies primes for reductions of elliptic curves on average. *LMS J. Comput. Math.*, 18(1):308–322, 2015.
24. Andrew V. Sutherland. Isogeny volcanoes. In *ANTS X—Proceedings of the Tenth Algorithmic Number Theory Symposium*, volume 1 of *Open Book Ser.*, pages 507–530. Math. Sci. Publ., Berkeley, CA, 2013.
25. Jacques Vélu. Isogénies entre courbes elliptiques. *C. R. Acad. Sci. Paris Sér. A-B*, 273:A238–A241, 1971.
26. John Voight. *Quaternion Algebras*. v.0.9.12, March 29, 2018.
27. William C. Waterhouse. Abelian varieties over finite fields. *Ann. Sci. École Norm. Sup. (4)*, 2:521–560, 1969.
28. Youngho Yoo, Reza Azarderakhsh, Amir Jalali, David Jao, and Vladimir Soukharev. A post-quantum digital signature scheme based on supersingular isogenies. In *Financial Cryptography and Data Security - 21st International Conference, FC 2017, Sliema, Malta, April 3–7, 2017, Revised Selected Papers*, pages 163–181, 2017.

Chabauty–Coleman Experiments for Genus 3 Hyperelliptic Curves

Jennifer S. Balakrishnan, Francesca Bianchi, Victoria Cantoral-Farfán, Mirela Çiperiani, and Anastassia Etropolski

Abstract We describe a computation of rational points on genus 3 hyperelliptic curves C defined over $\mathbb{Q}$ whose Jacobians have Mordell–Weil rank 1. Using the method of Chabauty and Coleman, we present and implement an algorithm in SageMath to compute the zero locus of two Coleman integrals and analyze the finite set of points cut out by the vanishing of these integrals. We run the algorithm on approximately 17,000 curves from a forthcoming database of genus 3 hyperelliptic curves and discuss some interesting examples where the zero set includes global points not found in $C(\mathbb{Q})$.

Keywords Chabauty-Coleman method · Rational points · Hyperelliptic curves

J. S. Balakrishnan
Department of Mathematics and Statistics, Boston University, 111 Cummington Mall, Boston, MA 02215, USA
e-mail: jbala@bu.edu

F. Bianchi
Mathematical Institute, University of Oxford, Andrew Wiles Building, Radcliffe Observatory Quarter, Woodstock Road, Oxford OX2 6GG, UK
e-mail: francesca.bianchi@maths.ox.ac.uk

V. Cantoral-Farfán
The Abdus Salam International Center for Theoretical Physics, Mathematics Section, 11 Strada Costiera, 34151 Trieste, Italy
e-mail: vcantora@ictp.it

M. Çiperiani (✉)
Department of Mathematics, The University of Texas at Austin, 1 University Station, C1200, Austin, TX 78712, USA
e-mail: mirela@math.utexas.edu

A. Etropolski
Department of Mathematics, Rice University MS 136, Houston, TX 77251, USA
e-mail: aetropolski@rice.edu

© The Author(s) and The Association for Women in Mathematics 2019
J. S. Balakrishnan et al. (eds.), *Research Directions in Number Theory*, Association for Women in Mathematics Series 19, https://doi.org/10.1007/978-3-030-19478-9_3

1 Introduction

Let C be a non-singular curve over $\mathbb{Q}$ of genus g. In the case where $g = 0$ or $g = 1$, C has extra structure given by the fact that if $C(\mathbb{Q})$ is non-empty, then C is rational (if $g = 0$) or C is an elliptic curve (if $g = 1$). In these cases, computing the set of rational points is either trivial by the Hasse principle or highly non-trivial in the case of elliptic curves. In the latter case the rational points form a finitely generated abelian group. Methods specific to this case exist for computing upper bounds on the rank of $C(\mathbb{Q})$, and the possibilities for the torsion subgroup of $C(\mathbb{Q})$ are completely understood by the work of Mazur [20, Theorem 8].

On the other hand, if $g \geq 2$, then C is of general type, and the Mordell conjecture, proved by Faltings in 1983 [16], implies that $C(\mathbb{Q})$ is finite. Our main motivation is to compute $C(\mathbb{Q})$ explicitly in this case. We will focus our attention on hyperelliptic curves of genus 3 such that the group of rational points of the Jacobian of C has Mordell–Weil rank $r = 1$. This falls into the special case where $r < g$ which was considered by Chabauty in 1941 [12], and techniques developed by Coleman in the 1980s allow us to use p-adic integration to bound, and often, in practice, actually compute the set of rational points [14, 15].

In addition to these methods, we will also use the algorithm of Balakrishnan, Bradshaw, and Kedlaya [3] and its implementation in SageMath [24] to explicitly compute the relevant Coleman integrals by computing analytic continuation of Frobenius on curves. Nonetheless, we note that the algorithms presented in this article (see Section 3) have not been implemented previously by other authors or carried out on a large collection of curves. (See, however, [7] for related work in genus 2.) Our code is available at [2].

We consider the case of genus 3 hyperelliptic curves for two reasons:

(1) When $g = 3$, we can impose the condition that $0 < r < g - 1$, i.e., $r = 1$, which, by a dimension argument, makes the method more effective. Indeed, in this case, the set $C(\mathbb{Q})$ is contained in the intersection of the zero sets of the integrals of two linearly independent regular 1-forms on the base change of C to $\mathbb{Q}_p$, where p is any odd prime of good reduction.
(2) When $g = 2$, the Jacobian of C is a surface, and its geometry and arithmetic is better understood. In particular, methods developed by Cassels and Flynn have been implemented by Stoll in Magma to make the computations needed much more efficient. More precisely, in this case, one can simplify the algorithm further by working with the quotient of the Jacobian by $\langle \pm 1 \rangle$, which is a quartic surface in $\mathbb{P}^3$, known as the Kummer surface. In order to make the search of rational points more effective, the Chabauty method can also be combined with the Mordell–Weil sieve, which uses information at different primes (see also [8]).

We begin with an overview of the Chabauty–Coleman method and explicit Coleman integration in Section 2. In Section 3, we present an algorithm to find a finite set of p-adic points containing the rational points of a hyperelliptic curve[1] $C/\mathbb{Q}$ of genus 3 that admits an odd model and whose Jacobian J has rank 1. We fix a prime p and work under the assumption that we know a $\mathbb{Q}$-rational point whose image in the Jacobian has infinite order (here the embedding of C into J is via a chosen basepoint ∞). Besides $\mathbb{Q}$-rational points, the output will include all points in $C(\mathbb{Q}_p)$ which are in the preimage of the p-adic closure of $J(\mathbb{Q})$ in $J(\mathbb{Q}_p)$.

We then proceed to run our code on a list of relevant curves taken from the forthcoming database of genus 3 hyperelliptic curves [6]. Our list consists of 16,977 curves, and we separately do a point search in Magma to find all $\mathbb{Q}$-rational points whose x-coordinates with respect to a fixed integral affine model have naive height at most 10^5 (cf. Section 4). Our Chabauty–Coleman computations then show that there are no $\mathbb{Q}$-rational points of larger height on any of these curves.

In some cases, our algorithm outputs points in $C(\mathbb{Q}_p) \setminus C(\mathbb{Q})$. Besides $\mathbb{Q}_p$-rational (but non-$\mathbb{Q}$-rational) Weierstrass points, on 75 curves we find that the local point is the localization of a point $P \in C(K)$, where K is a quadratic field in which the prime p splits. In all these cases, we are able to explain why these points appear in the zero locus that we are studying. The following three scenarios occur, and we discuss representative examples of each in Section 4:

- Case 1: It may happen that $[P - \infty]$ is a torsion point in the Jacobian (see Example 4.1). In this case, the integral of any 1-form would vanish between ∞ and P.
- Case 2: As in Example 4.2, it may happen that some multiple of the image of $[P - \infty]$ in the Jacobian actually belongs to $J(\mathbb{Q})$: the vanishing here follows by linearity in the endpoints of integration.
- Case 3: The Jacobian J may decompose over $\mathbb{Q}$ as a product of an elliptic curve and an abelian surface. Then if the subgroup H generated by $J(\mathbb{Q})$ and the point $[P - \infty]$ comes from the elliptic curve, the dimension of the p-adic closure of H in $J(\mathbb{Q}_p)$ must be equal to 1, even if $[P - \infty]$ is a point of infinite order (see Example 4.3).

We conclude in Section 4.2 by describing some follow-up computations in Cases 1 and 2 in light of Stoll's strong Chabauty conjecture.

Acknowledgements. The first author is supported in part by NSF grant DMS-1702196, the Clare Boothe Luce Professorship (Henry Luce Foundation), and Simons Foundation grant #550023. The second author is supported by EPSRC and by Balliol College through a Balliol Dervorguilla scholarship. The third author was supported by a Conacyt fellowship. The fourth author is supported by NSF grant DMS-1352598.

This project began at "WIN4: Women in Numbers 4," and we are grateful to the conference organizers for facilitating this collaboration. We further acknowledge

---

[1] We focus on hyperelliptic curves defined over $\mathbb{Q}$ because SageMath has an implementation of Coleman integration precisely for odd degree hyperelliptic curves given by a monic equation defined over $\mathbb{Z}_p$ for odd primes p of good reduction.

the hospitality and support provided by the Banff International Research Station. We thank the Simons Collaboration on Arithmetic Geometry, Number Theory, and Computation for providing computational resources, and we are grateful to Alexander Best, Raymond van Bommel, Bjorn Poonen, Michael Stoll, Andrew Sutherland, and Felipe Voloch for helpful conversations. Finally, we would also like to thank the anonymous referee for their careful work, which resulted in a number of improvements to the article.

2 The Chabauty–Coleman Method and Coleman Integration

In this section, we review the Chabauty–Coleman method, used to compute rational points in our main algorithm. For further details, see Section 3. We also give a brief overview of explicit Coleman integration on hyperelliptic curves.

2.1 Chabauty–Coleman Method

Let C be a smooth, projective curve over the rationals of genus at least 2. Approximately 40 years before Faltings' non-constructive proof of finiteness of $C(\mathbb{Q})$ [16], Chabauty considered the following setup. Let p be a prime and suppose that there exists $P \in C(\mathbb{Q})$, so the embedding

$$\iota_P : C \hookrightarrow J$$

$$Q \mapsto [Q - P]$$

maps $C(\mathbb{Q})$ in $J(\mathbb{Q})$.

Then let $\overline{J(\mathbb{Q})}$ denote the p-adic closure of $J(\mathbb{Q})$ and define

$$C(\mathbb{Q}_p) \cap \overline{J(\mathbb{Q})} := \iota_P(C(\mathbb{Q}_p)) \cap \overline{J(\mathbb{Q})}.$$

Chabauty proved the following case of Mordell's conjecture:

Theorem 2.1 ([12]). *Let $C/\mathbb{Q}$ be a curve of genus $g \geq 2$ such that the Mordell–Weil rank of the Jacobian J of C over $\mathbb{Q}$ is less than g, and let p be a prime. Then $C(\mathbb{Q}_p) \cap \overline{J(\mathbb{Q})}$ is finite.*

Chabauty's result was later reinterpreted and made effective by Coleman, who showed the following:

Theorem 2.2 ([14]). *Let C be as above and suppose that p is a prime of good reduction for C. If $p > 2g$, then*

$$\#C(\mathbb{Q}) \leq \#C(\mathbb{F}_p) + 2g - 2.$$

To obtain an explicit upper bound on the size of $C(\mathbb{Q}_p) \cap \overline{J(\mathbb{Q})}$, and hence of $C(\mathbb{Q})$, Coleman used his theory of p-adic integration on curves to construct p-adic integrals of 1-forms on $J(\mathbb{Q}_p)$ that vanish on $J(\mathbb{Q})$ and restrict them to $C(\mathbb{Q}_p)$. Here, we follow the exposition in [27] in defining the Coleman integral.

Let $\omega_J \in H^0(J_{\mathbb{Q}_p}, \Omega^1)$, and let λ_{ω_J} be the unique analytic homomorphism $\lambda_{\omega_J} : J(\mathbb{Q}_p) \to \mathbb{Q}_p$ such that its derivative $d(\lambda_{\omega_J})$ is exactly ω_J (cf. [27, §3 and §4] and [22, §4.1]). Consider the map

$$\iota^* : H^0(J_{\mathbb{Q}_p}, \Omega^1) \to H^0(C_{\mathbb{Q}_p}, \Omega^1)$$

induced by ι_P. Observe that ι^* is an isomorphism of vector spaces [21, Proposition 2.2] which is independent of the choice of $P \in C(\mathbb{Q}_p)$ [27, Lemma 4.2].

Define $\omega := \iota^*(\omega_J)$ to be the corresponding differential on C. On the Jacobian we have the natural pairing

$$\lambda : H^0(J_{\mathbb{Q}_p}, \Omega^1) \times J(\mathbb{Q}_p) \to \mathbb{Q}_p$$

$$(\omega_J, R) \mapsto \lambda_{\omega_J}(R) =: \int_0^R \omega_J.$$

Note that since λ_{ω_J} is a homomorphism, it vanishes on $J(\mathbb{Q}_p)_{\text{tors}}$. Now given $P, Q \in C(\mathbb{Q}_p)$ we define

$$\int_P^Q \omega := \int_0^{[Q-P]} \omega_J,$$

hence for a fixed point $P \in C(\mathbb{Q}_p)$ and $\omega \in H^0(C_{\mathbb{Q}_p}, \Omega^1)$ we get a function $\lambda_{\omega,P} : C(\mathbb{Q}_p) \to \mathbb{Q}_p$ with

$$\lambda_{\omega,P}(Q) := \int_P^Q \omega = \int_0^{[Q-P]} \omega_J = \lambda_{\omega_J}([Q - P]).$$

We now restrict to $g = g(C) = 3$ and $r = \text{rank } J(\mathbb{Q}) = 1$, in which case $g - r = 2$. The exposition below can be generalized whenever $r < g$. Let

$$\text{Ann}(J(\mathbb{Q})) := \left\{ \omega_J \in H^0(J_{\mathbb{Q}_p}, \Omega^1) : \lambda_{\omega_J}(R) = \int_0^R \omega_J = 0 \text{ for all } R \in J(\mathbb{Q}) \right\}.$$

This is a 2-dimensional $\mathbb{Q}_p$-vector space because there is an $\omega_J \in H^0(J_{\mathbb{Q}_p}, \Omega^1)$ such that λ_{ω_J} is non-trivial on $J(\mathbb{Q})$ since $J(\mathbb{Q})$ is not torsion (cf. [15, Proposition 3.1]). Hence there exist two linearly independent differentials $\alpha_J, \beta_J \in H^0(J_{\mathbb{Q}_p}, \Omega^1)$ such that

$$\lambda_{\alpha_J}(R) = \lambda_{\beta_J}(R) = 0 \quad \text{for all } R \in J(\mathbb{Q}).$$

Let D be a $\mathbb{Q}$-rational divisor on C of degree d, and consider the map $\iota_D : C \to J$ such that $Q \mapsto [dQ - D]$. Define

$$\lambda_{\omega,D}(Q) := \lambda_{\omega_J} \circ \iota_D(Q) = \lambda_{\omega_J}([dQ - D]).$$

Consider the set

$$Z := \{Q \in C(\mathbb{Q}_p) : \lambda_{\omega,D}(Q) = 0 \text{ for all } \omega \in \text{Ann}(J(\mathbb{Q}))\}$$

$$= \ker(\lambda_{\alpha,D}) \cap \ker(\lambda_{\beta,D}). \tag{2.1}$$

The upshot is that this set Z contains the rational points $C(\mathbb{Q})$ by construction. Moreover, while a priori we have defined Z in terms of D, it is actually independent of the choice [27, §1.6].

The above discussion indicates how we would handle the case when our hyperelliptic curve has an even degree model. However, since we restrict our attention to hyperelliptic curves C with an odd degree model, we are guaranteed a rational point $\infty \in C(\mathbb{Q})$ and we use $D = \infty$. Hence, we have two $\mathbb{Q}_p$-valued functions $\lambda_{\alpha,\infty}, \lambda_{\beta,\infty}$ on $C(\mathbb{Q}_p)$ whose common zeros capture the rational points of C.

2.2 Computing Coleman Integrals

In order to compute Z, we need a way to evaluate $\int_P^Q \omega$ for an arbitrary $\omega \in H^0(C_{\mathbb{Q}_p}, \Omega^1)$ and arbitrary $P, Q \in C(\mathbb{Q}_p)$. Suppose that p is a prime of good reduction for C and let $\overline{C}$ be the reduction of C modulo p, i.e., the special fiber of a minimal regular proper model of C over $\mathbb{Z}_p$. Then there exists a natural reduction map $C(\mathbb{Q}_p) \to \overline{C}(\mathbb{F}_p)$. Define a **residue disk** to be a fiber of the reduction map. To compute $\int_P^Q \omega$, we now consider two cases: either P and Q lie in the same residue disk or they do not.

2.2.1 Coleman Integral Within a Residue Disk

Let $P \in C(\mathbb{Q}_p)$. By a **local coordinate** for P we mean a rational function $t \in \mathbb{Q}_p(C)$ such that

(1) t is a uniformizer at P; and
(2) the reduction of t to a rational function for $\overline{C}$ is a uniformizer at $\overline{P}$.

Hence, a local coordinate t at P establishes a bijection

$$p\mathbb{Z}_p \leftrightarrow \{Q \in C(\mathbb{Q}_p) : \overline{Q} = \overline{P}\}$$

$$t \leftrightarrow (x(t), y(t)),$$

where $x(t), y(t)$ are Laurent series and $(x(0), y(0)) = P$.

Suppose now that $\omega \in H^0(C_{\mathbb{Q}_p}, \Omega^1)$ is normalized by an element of $\mathbb{Q}_p^\times$ in such a way that its reduction modulo p exists and is not identically zero. We will often use the fact that the expansion of ω in terms of t has the form $w(t)dt$, for some $w(t) \in \mathbb{Z}_p[[t]]$ converging on the entire residue disk. Hence, if $\overline{Q} = \overline{P}$, we can compute $\int_P^Q \omega$ by formally integrating a power series in the local coordinate t ([27, Lemma 7.2]). Such definite integrals are referred to as tiny integrals.

A local coordinate at a given point $P \in C(\mathbb{Q}_p)$ can be found using [1, Algorithms 2–4]. In particular, when $C : y^2 = F(x)$ with $F(x) \in \mathbb{Q}[x]$ of degree 7, we have

(1) If $y(\overline{P}) \neq 0$, then $x(t) = t + x(P)$ and $y(t)$ is the unique solution to $y^2 = F(x(t))$ such that $y(0) = y(P)$.
(2) If $y(\overline{P}) = 0$ and $\overline{P} \neq \infty$, then $y(t) = t + y(P)$ and $x(t)$ is the unique solution to $F(x) = y(t)^2$ such that $x(0) = x(P)$.
(3) If $\overline{P} = \infty$, one first finds $x(t) = t^{-2} + O(1)$ by solving $F(x) = \frac{x^6}{t^2}$. Then $y(t) = \frac{x(t)^3}{t}$.

In practice, in all three cases one can explicitly compute $(x(t), y(t))$ up to arbitrary p-adic and t-adic precision by Newton's method.

2.2.2 Coleman Integral Between Different Residue Disks

In our intended application of computing rational points, we will fix a basepoint as one endpoint of integration and consider the various Coleman integrals given by varying the other endpoint of integration over all residue disks. This makes it essential that the tiny integrals constructed in the previous section are consistent across the set of residue disks: in other words, we need a notion of analytic continuation between different residue disks.

Coleman solved this problem by using Frobenius to write down a unique "path" between different residue disks and presented a theory of p-adic line integration on curves [15] satisfying a number of natural properties, among them linearity in the integrand, additivity in endpoints, change of variables via rigid analytic maps (e.g., Frobenius), and the fundamental theorem of calculus. This was made algorithmic in [3] for *hyperelliptic* curves by solving a linear system induced by the action of Frobenius on Monsky-Washnitzer cohomology, with an implementation available in SageMath.

Thus given two points $P, Q \in C(\mathbb{Q}_p)$, one can compute the definite Coleman integral from P to Q as $\int_P^Q \omega$ directly via [3], as well as the Coleman integral from P to the residue disk of Q, by further computing a local coordinate t_Q at Q (such that $t_Q|_{t=0} = Q$), which gives

$$\int_P^{t_Q} \omega = \int_P^Q \omega + \int_Q^{t_Q} \omega = \int_P^Q \omega + \int_0^t \omega,$$

where $\int_P^Q \omega$ now plays the role of the constant of integration between different residue disks.

3 The Algorithm

We now specialize to our case of interest, where C is a genus 3 hyperelliptic curve given by an odd degree model, i.e.,

$$C: y^2 = F(x),$$

where $F(x) \in \mathbb{Q}[x]$ is monic of degree 7. We will further assume that the Jacobian J of C has Mordell–Weil rank 1 over $\mathbb{Q}$. Finally, we will assume that we have computed a point $P_0 \in C(\mathbb{Q})$ with the property that $[P_0 - \infty]$ is of infinite order in $J(\mathbb{Q})$. (This last assumption is straightforward to remove for a particular choice of curve, see §3.4.1.)

Fix an odd prime p of good reduction for C, denote by $\overline{C}$ the base change of C to $\mathbb{F}_p$, and let $C(\mathbb{Q})_{\text{known}}$ denote a list of known points in $C(\mathbb{Q})$. Given this input, the algorithm in this section returns the set Z of common zeros of $\lambda_{\alpha,\infty}$ and $\lambda_{\beta,\infty}$, as defined in Section 2.1, excluding the known rational points $C(\mathbb{Q})_{\text{known}}$.

3.1 Upper Bounds in Residue Disks

Define $\omega_i = (x^i/2y)dx$ for $i \in \{0, 1, 2\}$. These differentials form a basis for $H^0(C_{\mathbb{Q}_p}, \Omega^1)$. Let α and β be 1-forms in $H^0(C_{\mathbb{Q}_p}, \Omega^1)$ such that α_J and β_J form a basis for $\text{Ann}(J(\mathbb{Q}))$ and such that α and β are not identically zero modulo p. We may assume that we are in one of the following two situations:

(1) $\alpha = \omega_0$ and β is a $\mathbb{Z}_p$-linear combination of ω_1 and ω_2, or
(2) α is a $\mathbb{Z}_p$-linear combination of ω_0 and ω_1 and β is a $\mathbb{Z}_p$-linear combination of ω_0 and ω_2.

Let $f'(t)$ be the local expansion of α or β in the residue disk of a point $\overline{Q} \in \overline{C}(\mathbb{F}_p)$. Ultimately we want to compute the zeros of a particular antiderivative $f(t)$ lying in $p\mathbb{Z}_p$ up to a desired p-adic precision. In certain cases, we will be able to avoid this calculation by instead obtaining an upper bound for the number of zeros of $f(t)$ in $p\mathbb{Z}_p$ which we know to be sharp. To do this, we use the theory of Newton polygons for p-adic power series (see, e.g., [19, IV.4]).

Given $f(t) \in \mathbb{Q}_p[[t]]$ such that $f'(t) \in \mathbb{Z}_p[[t]]$, let $\overline{f'}(t) = f'(t) \mod p$ and define

$$m_f := \text{ord}_{t=0}\overline{f'}(t). \tag{3.1}$$

The following result is [22, Lemma 5.1] and can be viewed as a corollary of the p-adic Weierstrass Preparation Theorem [19, Ch. IV, §4, Theorem 14].

Lemma 3.1. Let $f(t) \in \mathbb{Q}_p[[t]]$ such that $f'(t) \in \mathbb{Z}_p[[t]]$. If $m_f < p - 2$, then the number of roots of f in $p\mathbb{Z}_p$ is less than or equal to $m_f + 1$.

Remark 3.2. Note that if $p > 2g = 6$ and $f'(t)$ is the local expansion of a regular 1-form, then the Riemann-Roch Theorem implies that $m_f \leq 4$ and hence the condition $m_f < p - 2$ of Lemma 3.1 is always satisfied (cf. [22, Theorem 5.3]).

The following lemmas give a refinement of this result for our particular choice of f. We refer to a point of C or $\overline{C}$ as a **Weierstrass point** if it is fixed by the hyperelliptic involution and as a **finite Weierstrass point** if it is furthermore different from the point at infinity.

Lemma 3.3. *Let $f'(t)$ be the local expansion of α or β in the residue disk of a point $\overline{Q} \in \overline{C}(\mathbb{F}_p)$. If $\overline{Q}$ is non-Weierstrass, then*

$$
m_f \leq \begin{cases} 1 & \text{if } x(\overline{Q}) \neq 0, \\ 2 & \text{otherwise.} \end{cases}
$$

Moreover, the minimum of the orders of vanishing of α and β at $\overline{Q}$ is less than or equal to 1 for all non-Weierstrass $\overline{Q}$.

Proof. By construction, the differential f' is a linear combination of two of the differentials $\omega_i = (x^i/2y)dx$, $i = 0, 1, 2$, and f' is non-trivial modulo p. The assumption that $\overline{Q}$ is non-Weierstrass implies that $t = x - x(Q)$ is a local coordinate, where Q is any lift of $\overline{Q}$ to characteristic zero. Write $f' = ((Ax^i + Bx^j)/2y)dx$, where $A, B \in \mathbb{Z}_p, i, j \in \{0, 1, 2\}, i < j$. Then in local coordinates we have

$$
f'(t) = \frac{A(t + x(Q))^i + B(t + x(Q))^j}{2y(t)} dt,
$$

where $y(t)$ has no zeros or poles in the residue disk. Since $A(t + x(Q))^i + B(t + x(Q))^j$ is a polynomial of degree less than or equal to 2 in t, the first part of the first claim is proved. Furthermore, the polynomial has a double root modulo p at $\overline{Q}$ if and only if $A \equiv 0 \bmod p$, $j = 2$, $B \not\equiv 0$, and $x(\overline{Q}) = 0$, i.e., if and only if $f' \equiv \frac{x^2}{2y}dx$ (up to rescaling) and $x(\overline{Q}) = 0$. The last statement also follows, since by construction α is a linear combination of ω_0 and ω_1 (see the beginning of § 3.1). $\square$

Proposition 3.4. *Let p be a prime greater than or equal to 5 of good reduction for C. Let $\overline{Q} \in \overline{C}(\mathbb{F}_p)$ be a non-Weierstrass point. Then the set*

$$
\left\{ P \in C(\mathbb{Q}_p) : \overline{P} = \overline{Q} \text{ and } \int_{\infty}^{P} \alpha = \int_{\infty}^{P} \beta = 0 \right\}
$$

has size less than or equal to 2.

Proof. Follows from Lemma 3.1 and Lemma 3.3. $\square$

Lemma 3.5. *Let $f'(t)$ be the local expansion of α or β in the residue disk of the point $\overline{\infty} \in \overline{C}(\mathbb{F}_p)$. Then $m_f \in \{0, 2, 4\}$. In particular, the minimum of the orders of vanishing of α and β at $\overline{\infty}$ is less than or equal to 2.*

Proof. We may take $x(t) = t^{-2} + O(1)$, $y(t) = t^{-7} + O(t^{-5})$ (cf. [1, Algorithm 4]). Then $x^i dx / 2y$ has a zero of order $4 - 2i$ at $t = 0$. □

Proposition 3.6. *Let p be a prime greater than or equal to 5 of good reduction for C. Then the set*

$$\left\{ P \in C(\mathbb{Q}_p) : \overline{P} = \overline{\infty} \text{ and } \int_{\infty}^{P} \alpha = \int_{\infty}^{P} \beta = 0 \right\}$$

has size less than or equal to 3. In particular, there are at most two points different from the point at infinity and reducing to it modulo p in the above set.

Lemma 3.7. *Let $f'(t)$ be the local expansion of α or β in the residue disk of a point $\overline{Q} \in \overline{C}(\mathbb{F}_p)$ (with the notation of the algorithm). If $\overline{Q}$ is a finite Weierstrass point, then*

$$m_f \in \begin{cases} \{0, 2\} & \text{if } x(\overline{Q}) \neq 0, \\ \{0, 2, 4\} & \text{else.} \end{cases}$$

Moreover, the minimum of the orders of vanishing of α and β at $\overline{Q}$ is less than or equal to 2.

Proof. In this case we may take $y = t$ and solve for x using $y^2 = F(x)$. In particular, we have that $x(t) = x(Q) + \frac{t^2}{F'(x(Q))} + O(t^4) \bmod p$ (cf. [1, Algorithm 3]). Therefore, $dx/2y$ has no zero or pole at $t = 0$ and $x^i dx / 2y$ has either no zero or pole or a zero of order $2i$ if $\overline{Q} = (0, 0)$. Now consider

$$f'(t) = \left(A \left(x(Q) + \frac{t^2}{F'(x(Q))} + O(t^4) \right)^i + B \left(x(Q) + \frac{t^2}{F'(x(Q))} + O(t^4) \right)^j \right) u(t) dt,$$

where $u(t)$ is a unit power series and A is non-zero modulo p. For any choice of $i, j \in \{0, 1, 2\}, i < j$, it can be verified that $m_f \in \{0, 2\}$ by distinguishing between the cases $x(\overline{Q}) = 0$ or $x(\overline{Q}) \neq 0$. If $A \equiv 0 \bmod p$ when $i = 0$ or 1 and $j = 2$, then m_f equals 4 if $x(\overline{Q}) = 0$. However, by construction, both α and β cannot be of this form. □

Proposition 3.8. *Let p be a prime greater than or equal to 5 of good reduction for C. Let $\overline{Q} \in \overline{C}(\mathbb{F}_p)$ be a finite Weierstrass point. Then the set*

$$\left\{ P \in C(\mathbb{Q}_p) : \overline{P} = \overline{Q} \text{ and } \int_{\infty}^{P} \alpha = \int_{\infty}^{P} \beta = 0 \right\}$$

has size less than or equal to 3.

3.2 Roots of p-Adic Power Series

Let $f'(t)$ be the local expansion of α (resp. β) in a residue disk, and let $f(t)$ be an antiderivative of $f'(t)$ whose constant term is either zero or the Coleman integral of α (resp. β) between ∞ and a $\mathbb{Q}_p$-rational point on C. To provably determine the roots of $f(t)$ lying in a residue disk up to a desired p-adic precision, we need to do the following:

- truncate at a p-adic precision p^N that is able to detect all the roots (up to $O(p^n)$, where $n = \lfloor (N-k)/2 \rfloor$, see Proposition 3.11);
- determine M such that to compute a root up to $O(p^n)$, we only need to consider the power series up to $O(t^M)$, where the coefficient of t^i has p-adic valuation greater than or equal to $2n$ for all $i \geq M$ if the roots are simple and f is suitably normalized (i.e., $f \in \mathbb{Z}_p[[t]] \setminus p\mathbb{Z}_p[[t]]$).

Write $f(t) = f_M(t) + O(t^M)$, where M is an integer greater than or equal to $m_f + 2$ and $f_M(t)$ is a polynomial of degree less than or equal to $M - 1$. Then $f(t)$ and $f_M(t)$ have the same number of roots in $\mathbb{C}_p$ of p-adic valuation greater than or equal to 1, as can be deduced from the same considerations on the Newton polygon of $f(t)$ which imply Lemma 3.1 (for more details, see the proof of [22, Lemma 5.1]). We are interested in the zeros of $f(pt)$ in $\mathbb{Z}_p$. Note that

$$f(pt) - f_M(pt) \in O(p^n, t^M) \text{ for some } n.$$

Hence, $f(pt)$ and $f_M(pt)$ have exactly the same zeros in $\mathbb{Z}/p^{n-k}\mathbb{Z}$ (including multiplicities), where k is the minimal valuation of the coefficients of $f(pt)$. Furthermore, if a zero of $p^{-k} f(pt)$ (and $p^{-k} f_M(pt)$) modulo p^{n-k} is simple (i.e., its derivative is non-zero modulo $p^{\lceil (n-k)/2 \rceil}$), then it lifts to a root of $f(pt)$ in $\mathbb{Z}_p$ by an inductive application of Hensel's lemma.

To compute a suitable choice of M, we require two more lemmas.

Lemma 3.9. *Let* $\omega_i = (x^i/2y)dx$ *for some* $i \in \{0, 1, 2\}$, *and* $\lambda_i = \int_\infty^{P_0} \omega_i$. *If* $\overline{[P_0 - \infty]} \in J(\mathbb{F}_p)$ *has order prime to* p, *then* $\mathrm{ord}_p(\lambda_i) \geq 1$. *In particular, this holds if* p *does not divide* $\#J(\mathbb{F}_p)$.

Proof. Let n be the order of the reduction of $[P_0 - \infty]$ modulo p. Then $Q = n[P_0 - \infty] \in J_1(\mathbb{Q}_p)$, the kernel of reduction at p, and we have $\int_\infty^{P_0} \omega_i = \frac{1}{n} \int_0^Q \omega_{J,i}$, where $\iota^*(\omega_{J,i}) = \omega_i$. Now $\int_0^Q \omega_{J,i}$ can be computed by writing $\omega_{J,i}$ as a power series in $\mathbb{Z}_p[[z_1, z_2, z_3]]$ where z_1, z_2, z_3 is a local coordinate system for $J_1(\mathbb{Q}_p)$ around 0, formally integrating and evaluating at $z_1(Q), z_2(Q), z_3(Q)$. $\square$

Lemma 3.10. *We have* $f(pt) = \sum_{i=0}^\infty b_i t^i = \sum_{j=0}^\infty \frac{a_j p^{j+1}}{j+1} t^{j+1} + c$, *where* $c \in \mathbb{Q}_p$, $a_j \in \mathbb{Z}_p$ *for all* $j \geq 0$. *Therefore for all* $i \geq 1$, $\mathrm{ord}_p(b_i) \geq i - \mathrm{ord}_p(i)$. *Furthermore if* $p^2 \nmid \#J(\mathbb{F}_p)$, *then* $c \in \mathbb{Z}_p$.

Proof. The first assertion is clear. For the latter, recall that c is either 0 or of the form $\int_\infty^Q \gamma$, for some $Q \in C(\mathbb{Q}_p)$ and $\gamma \in \{\alpha, \beta\}$. The proof is then similar to Lemma 3.9. □

By Lemma 3.10, we know that $f(pt)$ has coefficients in $\mathbb{Z}_p$, except possibly when $p^2 | \#J(\mathbb{F}_p)$. Let k be the minimum of the valuations of the coefficients of $f(pt)$. Note that, since $f'(t)$ mod p has order of vanishing equal to m_f, if $m_f < p - 2$, it follows that the valuation of the coefficient of t^{m_f+1} in $f(pt)$ is precisely $m_f + 1$. Therefore $k \le m_f + 1$. Furthermore, for $i > m_f + 1$, we have $\mathrm{ord}_p(b_i) \ge i - \mathrm{ord}_p(i) > i - (i - m_f - 1) = m_f + 1$.

Proposition 3.11. *Let $f(t)$ be an antiderivative of α or β, let $m_f < p - 2$, and let k be the minimal valuation of the coefficients of $f(pt)$. Fix an integer N such that $m_f + 2 \le N \le p^p - p$. Let ap^e be the smallest integer greater than or equal to N with $p \nmid a$ and $e \ge 1$, and set*

$$M = \begin{cases} ap^e + 1 & \text{if } ap^e - e < N, \\ N & \text{otherwise.} \end{cases}$$

Then each simple root of $p^{-k} f_M(pt)$ in $\mathbb{Z}/p^{N-k}\mathbb{Z}$ equals the approximation modulo $p^{\lfloor (N-k)/2 \rfloor}$ of a root of $f(pt)$. Furthermore, if all such roots are simple, then these are all the roots of $f(pt)$ in $\mathbb{Z}_p$.

Proof. It suffices to show that

$$\mathrm{ord}_p(b_i) \ge N \tag{3.2}$$

for all $i \ge M$. Indeed, if that is the case then

$$p^{-k} f_M(pt) = p^{-k} f(pt) \text{ modulo } p^{N-k} \quad \text{for all } t \in \mathbb{Z}_p.$$

Thus, if $t_0 \in \mathbb{Z}_p$ is a root of $f(pt)$, then t_0 modulo p^{N-k} is a root of $p^{-k} f_M(pt)$ modulo p^{N-k}. Conversely, if t_0 is a simple root of $p^{-k} f_M(pt)$ modulo p^{N-k}, then by Hensel's lemma this lifts to a root in $\mathbb{Z}_p$ of $f_M(pt)$ congruent to t_0 modulo $p^{\lfloor (N-k)/2 \rfloor}$.

Since $M \ge N$, the statement (3.2) is clear for $p \nmid i$ by Lemma 3.10. Now suppose $p | i$ for some $i \ge M$. Hence, $i = bp^r$, where $p \nmid b$ and $r \ge 1$, and $bp^r \ge M \ge N$. Then by the definition of ap^e, we know that

$$bp^r \ge ap^e \quad \text{and} \quad 0 \le ap^e - N < p.$$

We now have two cases to consider:

Case 1: Assume that $bp^r = ap^e$. It follows that $M = N$ which in turn implies that $ap^e - e \ge N$. Then since $(a, e) = (b, r)$, we have that

$$\mathrm{ord}_p(b_i) \ge bp^r - r \ge N.$$

Case 2: Assume that $bp^r > ap^e$. It follows that $bp^r - ap^e \geq p$. Thus

$$\text{ord}_p(b_i) \geq bp^r - r = (bp^r - ap^e) + (ap^e - r).$$

So if $\text{ord}_p(b_i) < N$, then $r > (bp^r - ap^e) + (ap^e - N) \geq p$ and hence $p^p - p \leq bp^r - r < N$, contradicting our assumption on N.   □

Remark 3.12. In order to apply Proposition 3.11 we need to meet the condition $m_f < p - 2$; assuming that $p > 2g$ guarantees that this is always the case, as a consequence of the Riemann-Roch Theorem (see Remark 3.2). Furthermore, in the case when $p > 2g$, the hypothesis on N of Proposition 3.11 is always met, since $m_f + 2 < p < p^p - p$.

3.3 Outline of the Algorithm

We retain the notation of the beginning of Section 3. The algorithm will always work if $p \geq 7$ and may or may not work if $p = 3$ or 5 (see Remark 3.2 and the comments in the main steps of the algorithm below). We now list the input and output of our algorithm followed by its main steps.

Input:

- C: a hyperelliptic curve of genus 3 over $\mathbb{Q}$ given by a model $y^2 = F(x)$, where $F \in \mathbb{Q}[x]$ is monic of degree 7, such that its Jacobian J has rank 1;
- p: an odd prime of good reduction for C not dividing the leading coefficient of F and $p \geq 7$;
- P_0: a point in $C(\mathbb{Q})$ such that $[P_0 - \infty] \in J(\mathbb{Q})$ has infinite order;
- $C(\mathbb{Q})_{\text{known}}$: a list of all known rational points on $C(\mathbb{Q})$;
- a p-adic precision N.

Output: The set $Z \subseteq C(\mathbb{Q}_p)$ defined in (2.1) modulo the action of the hyperelliptic involution. In our code, this set is split into the following:

- a list of points of Z which can be recognized as points in $C(\mathbb{Q}) \setminus C(\mathbb{Q})_{\text{known}}$ up to the hyperelliptic involution;
- a list of points $P \in Z$ such that $[P - \infty] \in J(\mathbb{Q}_p)_{\text{tors}}$, up to the hyperelliptic involution. Here, if P is not 2-torsion and is the localization of a point defined over a quadratic extension of $K/\mathbb{Q}$, then the coordinates in K are given as the corresponding minimal polynomials over $\mathbb{Q}$; in all other cases, the p-adic expansion of P is returned.
- a list of all remaining points $P \in Z$ (as above, if P is the localization of a point defined over a quadratic extension of $K/\mathbb{Q}$, then the coordinates in K are given as the corresponding minimal polynomials over $\mathbb{Q}$; in all other cases, the p-adic expansion of P is returned).

Main steps of the algorithm:

(1) *A basis for the annihilator.*
    For each basis differential $\omega_i = (x^i/2y)dx$ $(i = 0, 1, 2)$, compute

$$\lambda_i = \int_\infty^{P_0} \omega_i \quad \text{modulo } p^n,$$

where n may be less than N if some precision is lost in computing the Coleman integrals. Set $k_{ij} := \min\{\text{ord}_p(\lambda_i), \text{ord}_p(\lambda_j)\}$ and

$$(\alpha, \beta) = \begin{cases} \left(\omega_0, \ p^{-k_{12}}(\lambda_1\omega_2 - \lambda_2\omega_1)\right) & \text{if } \lambda_0 = 0, \\ \left(p^{-k_{01}}(\lambda_0\omega_1 - \lambda_1\omega_0), \ p^{-k_{02}}(\lambda_0\omega_2 - \lambda_2\omega_0)\right) & \text{else.} \end{cases}$$

In either case, α and β are reductions modulo $p^{n'}$ of the pullback ι^* of a basis for the annihilator of $J(\mathbb{Q})$, where

$$n' = \begin{cases} n - k_{12} & \text{if } \lambda_0 = 0, \\ n - \max\{k_{01}, k_{02}\} & \text{else.} \end{cases}$$

By Lemma 3.9, $n' \leq n - 1$ if $p \nmid \#J(\mathbb{F}_p)$. If $n' \geq 6$ we are guaranteed to be able to carry out all computations in the next steps when $p \geq 7$.

(2) *Ruling out residue disks.* Observe that we only need to consider residue disks up to the hyperelliptic involution.
    Reduce α and β modulo p. For each $\overline{P} \in \overline{C}(\mathbb{F}_p)$, expand α and β in a local coordinate s around $\overline{P}$, calculate the orders of vanishing of α and β at $s = 0$, and let $m(\overline{P})$ denote their minimum. Note that $m(\overline{P}) \leq 2$ by Lemmas 3.3, 3.5, and 3.7, and hence it suffices to compute $\alpha(s)$ and $\beta(s)$ up to $O(s^2)$ to find $m(\overline{P})$.
    If $m(\overline{P}) + 1$ equals the number of $\mathbb{Q}$-rational points in $C(\mathbb{Q})_{\text{known}}$ reducing to $\overline{P}$ modulo p and $m(\overline{P}) < p - 2$, then by Lemma 3.1 the set $C(\mathbb{Q})_{\text{known}}$ contains all $\mathbb{Q}$-rational points in the residue disk of $\overline{P}$. Otherwise, proceed to the next step.

(3) *Searching for the remaining disks.*
    If, for a given point $\overline{P} \in \overline{C}(\mathbb{F}_p)$, the number of $\mathbb{Q}$-rational points in $C(\mathbb{Q})_{\text{known}}$ reducing to $\overline{P}$ modulo p is strictly smaller that $m(\overline{P}) + 1$, then we need to compute the set of $\mathbb{Q}_p$-rational points P reducing to $\overline{P}$ such that $\int_Q^P \alpha = \int_Q^P \beta = 0$ for a (any) rational point Q. For computational convenience we distinguish between two cases:

(i) If there exists $P \in C(\mathbb{Q})_{\text{known}}$ reducing to $\overline{P}$, let t be a uniformizer at P. Then expand α and β in t and formally integrate to obtain two power series $f(t), g(t)$, which parameterize the integrals of α and β between P and any other point in the residue disk.

(ii) If we do not know any $\mathbb{Q}$-rational point in the residue disk of $\overline{P}$, then we may assume that $\overline{P} \neq \overline{\infty}$ and hence write $\overline{P} = (\overline{x_0}, \overline{y_0})$. If $\overline{y_0} = 0$, let $P = (x_0, 0)$, where x_0 is the Hensel lift of $\overline{x_0}$ to a root of $f(x)$. Otherwise, if $\overline{P}$ is not a Weierstrass point, we take $P = (x_0, y_0)$, where x_0 is any lift to $\mathbb{Z}_p$ of $\overline{x_0}$ (the Teichmüller lift of $\overline{x_0}$ is a particularly convenient choice for x_0, as it simplifies some of the underlying Coleman integration computations [3]) and y_0 is obtained from $\overline{y_0}$ using Hensel's Lemma on $y^2 = F(x_0)$. Let $\tilde{f}(t)$ and $\tilde{g}(t)$ be the integrals between P and any other point reducing to $\overline{P}$ in terms of a local parameter t at P. Then write $f(t) = \tilde{f}(t) + \int_{\infty}^{P} \alpha$ and $g(t) = \tilde{g}(t) + \int_{\infty}^{P} \beta$.

Recall that in (1), we have computed the coefficients of the ω_i in α and β modulo n'. To provably compute the set of common zeros, we require that each common root is a simple zero of at least one of f and g at our working precision. Assume without loss of generality that f has only simple roots. The t-adic precision we should compute $f(t)$ to in order to find provably correct approximations of its simple zeros is determined by Proposition 3.11. In practice, we truncate $f(t)$ at $O(t^M)$, where $M = n'$, unless the smallest multiple r of p greater than or equal to n' satisfies $r - \mathrm{ord}_p(r) < n'$, in which case take $M = r + 1$. For the p-adic precision, the coefficients are computed modulo $p^{n'}$. Then the simple roots are correct[2] up to $O(p^{\lfloor (n'-k)/2 \rfloor})$, where k is the minimal valuation of the coefficients of $f(pt)$ (by an argument analogous to the discussion preceding Proposition 3.11). To compute the roots we use the function `polrootspadic` implemented in PARI/GP. Finally, we take the list of roots which lie in $p\mathbb{Z}_p$ and check whether they are also roots of g.

If $p = 3$ or $p = 5$ and the order of vanishing of $f(t)$ or $g(t)$ modulo p is greater than or equal to $p - 2$, then we cannot provably find the zeros of $f(pt)$ and $g(pt)$. Currently the algorithm assumes that $p \geq 7$ to avoid these pitfalls.

(4) *Identifying the remaining disks.*

Once we have found the common zeros of $f(pt)$ and $g(pt)$, we recover the corresponding $\mathbb{Q}_p$-rational points that do not come from points in $C(\mathbb{Q})_{\mathrm{known}}$. We now have the output set that we will now break into sublists.

If we fail to recognize a point Q as $\mathbb{Q}$-rational, we can check whether the integral between ∞ and Q of any non-zero differential γ not in the span of α and β also vanishes. If this is the case, the point $[Q - \infty] \in J(\mathbb{Q}_p)$ is torsion (cf. [15, Proposition 3.1]) and the analogue of the Lutz–Nagell Theorem for hyperelliptic curves [17, Theorem 3], combined with reductions modulo primes, provides an effective procedure to determine which points of $C(\mathbb{Q})$ give rise to torsion points in $J(\mathbb{Q})$ and thus to verify whether Q is $\mathbb{Q}$-rational or not. Furthermore, by increasing the degree in `algdep`, we may even try to identify

---

[2]Note that when the derivative of $p^{-k} f_M(pt)$ at the root is non-zero modulo p which happens quite often, then the roots that we find are in fact correct modulo $p^{n'-k}$.

the number field over which the coordinates of Q are defined[3]. This may require high p-adic precision; however, it was possible for every curve we considered.

If the integral of the differential γ is non-zero, and we have not recognized Q as a $\mathbb{Q}$-rational point, we can still check whether the point Q is defined over some number field K. For instance, $[Q - \infty]$ could equal a point in $J(\mathbb{Q})$ plus some torsion element in $J(K)$, with $[K : \mathbb{Q}] > 1$ (see Example 4.1).

3.4 Generalizations of the Algorithm

3.4.1 What If We Do Not Know $P_0 \in C(\mathbb{Q})$ Such That $[P_0 - \infty] \notin J(\mathbb{Q})_{\text{tors}}$?

The hyperelliptic curve we input in the algorithm is assumed to have rank 1. Calculation of the rank is attempted by Magma [4] by working out both an upper bound and a lower bound, the former coming from computation of the 2-Selmer group and the latter from an explicit search for linearly independent points on the Jacobian. The success of the rank computation relies on the two bounds being equal. In particular, if we suppose that we know provably that the rank of the Jacobian is one, we may as well assume that we know a point $Q \in J(\mathbb{Q})$ of infinite order and a divisor E on C representing it. Then we may proceed as follows. The first task is to write $Q = [E]$ in the form $[D - d\infty]$, where D is an effective $\mathbb{Q}$-rational divisor. In order to achieve this, we follow step-by-step the proof of [26, Corollary 4.14]. That is, we compute the dimension of $\mathcal{L}(E + n\infty)$ for $n = 0, 1, 2, \ldots$ (here $\mathcal{L}(E + n\infty)$ denotes the Riemann-Roch space of $E + n\infty$), until we find the smallest $n = m$ for which the dimension is 1. Then $[D - d\infty] = [E + \text{div}(\phi)]$, where ϕ generates $\mathcal{L}(E + m\infty)$. By [26, Lemma 4.17], D is then the unique $\mathbb{Q}$-rational divisor in general position and of degree less than or equal to $g = 3$ such that Q can be represented in the form $[D - d\infty]$.

Let K be the smallest Galois extension of $\mathbb{Q}$ over which the support of D is defined. Furthermore, let p be a prime of good reduction for C that splits completely in $K/\mathbb{Q}$. Then K can be realized as a subfield of $\mathbb{Q}_p$ and hence the support of D can be seen as lying in $C(\mathbb{Q}_p)$. Write $D = \sum_{i=1}^{d} P_i$ (some P_i possibly being equal). Then we may proceed exactly as before, just replacing λ_i in (1) by

$$\lambda_i = \sum_{i=1}^{d} \int_{\infty}^{P_i} \omega_i.$$

---

[3]The hyperelliptic curve C is defined over $\mathbb{Q}$. Thus the fact that $[Q - \infty] \in J(\mathbb{Q}_p)_{\text{tors}}$ forces Q to have coordinates in $\overline{\mathbb{Q}} \cap \mathbb{Q}_p$.

3.4.2 Even Degree Model

The algorithm relies heavily on computations of Coleman integrals, for which one needs the hyperelliptic curve considered to have a model of the form $y^2 = F(x)$, where $F(x)$ is monic. In particular, if one were to work with an even degree model, the two points at infinity would necessarily be defined over $\mathbb{Q}$ [26]. Therefore we could proceed as in the odd degree case with the single point at infinity being replaced by one of these two points. If $F(x)$ is not monic, the issue is that Coleman integration is not implemented in SageMath, though an implementation is available in Magma [10, 11]. Hence, general even models could be handled by computing the set of local points Z as defined in (2.1), using the sum of the two points at infinity as the divisor D.

3.4.3 Other Ranks and Genera

Our assumptions on g and r are somewhat arbitrary. With minor modifications, our code can be used in more general cases, provided that $0 < r < g$.

4 Curve Analysis

Once we implemented in SageMath the algorithm described in the previous section, we ran it over $16,977$ hyperelliptic curves of genus $g = 3$ satisfying the following properties:

(1) the curve admits an odd degree model over $\mathbb{Q}$;
(2) the Jacobian of the curve has Mordell–Weil rank equal to 1;
(3) there is a $\mathbb{Q}$-rational point P_0 such that $[P_0 - \infty]$ has infinite order in $J(\mathbb{Q})$.

In order to obtain those curves, we sorted the $67,879$ genus 3 hyperelliptic curves from a forthcoming database of genus 3 curves over $\mathbb{Q}$ [6]. (For each hyperelliptic curve, this database has its absolute minimal discriminant, the conductor of its Jacobian, and an equation giving a minimal model for the curve.) Out of these, $19,254$ curves satisfy conditions (1) and (2).

Running our code for the $16,977$ curves for which we could further find a P_0 as in (3), we found 75 curves where the zero set Z contains something other than the rational points we had already computed and Weierstrass points defined over $\mathbb{Q}_p$ for our chosen prime p. Note that in all $16,977$ computations, the prime p used was the smallest prime greater than $2g = 6$ which divided neither the discriminant nor the leading coefficient of the hyperelliptic polynomial defining the curve.

Let C be one of these 75 curves, let W denote the set of Weierstrass points in $C(\mathbb{Q}_p) \setminus C(\mathbb{Q})$, and let $P \in Z \setminus (C(\mathbb{Q}) \cup W)$. In all cases, we identified P as a point defined over a quadratic extension K of $\mathbb{Q}$ in which p splits. Even so, these 75 curves split up into 3 distinct cases:

(1) $[P - \infty] \in J(K)_{\text{tors}}$;
(2) $[P - \infty] \notin J(K)_{\text{tors}}$ but $n[P - \infty] \in J(\mathbb{Q})$ for some positive integer n;
(3) $\langle J(\mathbb{Q}), [P - \infty] \rangle_{\mathbb{Z}}$ is a rank 2 subgroup of $J(K)$.

In Cases (1) and (2), it is clear why $P \in Z$. On the other hand, justifying Case (3) requires investigating more closely the geometry of the Jacobian of the curve, as is carried out in detail in Example 4.3.

4.1 Examples

For each of the curves below we give a list of known rational points, which are all of the rational points up to a height[4] of 10^5. Following the algorithm outlined in § 3.3, we produce the set Z of local points for the prime $p = 7$ or $p = 11$, which in each case returns no new $\mathbb{Q}$-rational points, hence concluding that the set of known rational points $C(\mathbb{Q})_{\text{known}}$ is all of $C(\mathbb{Q})$. In each of the examples below, however, Z contains a $\mathbb{Q}_p$-point which is not a Weierstrass point and falls into one of the cases outlined above.

Example 4.1. Consider the hyperelliptic curve

$$y^2 + x^3 y = x^7 + 2x^6 - 2x^5 - 9x^4 - 4x^3 + 8x^2 + 8x + 2$$

(given above by a minimal model) which has absolute minimal discriminant 544256 and whose Jacobian has conductor $544256 = 2^9 \cdot 1063$. We work with an odd degree model

$$C : y^2 = 4x^7 + 9x^6 - 8x^5 - 36x^4 - 16x^3 + 32x^2 + 32x + 8,$$

which has the following five known rational points:

$$C(\mathbb{Q})_{\text{known}} = \{\infty, (-1, -1), (-1, 1), (1, -5), (1, 5)\}.$$

---

[4]Our computations show that the $\mathbb{Q}$-rational points of highest absolute logarithmic height (with respect to an odd degree model) on a curve among the $16{,}977$ hyperelliptic curves that we considered are

$$\left(-\frac{49}{18}, -\frac{339563}{11664} \right), \left(-\frac{49}{18}, -\frac{1600445}{52488} \right)$$

on the hyperelliptic curve

$$C : y^2 + (x^4 + x^2 + x)y = x^7 - x^6 - 5x^5 + 5x^3 - 3x^2 - x,$$

which has absolute minimal discriminant 5326597 and whose Jacobian has conductor 5326597.

Running the code on C together with the prime $p = 7$ and the point $P_0 = (-1, -1)$, we find that

$$Z = C(\mathbb{Q})_{\text{known}} \cup W \cup \{(0, \pm 2\sqrt{2})\},$$

where the set W of non-$\mathbb{Q}$-rational Weierstrass points has size 3. Moreover, the points $[(0, \pm 2\sqrt{2}) - \infty] \in J(\mathbb{Q}(\sqrt{2}))$ have order 12.

Example 4.2. Consider the hyperelliptic curve

$$y^2 + (x^4 + 1)y = 2x^3 + 2x^2 + x$$

(given above by a minimal model) which has absolute minimal discriminant 48519 and whose Jacobian J has conductor $48519 = 3^4 \cdot 599$. We work with an odd degree model

$$C : y^2 = -4x^7 + 24x^6 - 56x^5 + 72x^4 - 56x^3 + 28x^2 - 8x + 1.$$

This curve has the following five known rational points:

$$C(\mathbb{Q})_{\text{known}} = \{\infty, (0, -1), (0, 1), (1, -1), (1, 1)\}.$$

For this example we will give a more detailed outline of the algorithm. Following Section 2.1 we know that there exist functions $\lambda_{\alpha,\infty}, \lambda_{\beta,\infty}$ on $C(\mathbb{Q}_p)$, corresponding to differentials $\alpha, \beta \in H^0(C_{\mathbb{Q}_p}, \Omega^1)$. These two functions vanish on the rational points of C, i.e.,

$$C(\mathbb{Q}) \subseteq \ker(\lambda_{\alpha,\infty}) \cap \ker(\lambda_{\beta,\infty}) = Z.$$

We would like to know whether this zero set Z contains anything other than the $\mathbb{Q}$-rational points on C.

If we take $p = 11$, we find that $Z = C(\mathbb{Q})_{\text{known}} \cup W$. Observe that 11 is inert in $K = \mathbb{Q}(\sqrt{-3})$.

If we take $p = 7$, which splits in $K = \mathbb{Q}(\sqrt{-3})$, we find that there are four points defined over $K = \mathbb{Q}(\sqrt{-3})$ that appear in Z. Up to hyperelliptic involution, we have

$$\left\{ \left((1 + \sqrt{-3})/2, \sqrt{-3} \right), \left((1 - \sqrt{-3})/2, \sqrt{-3} \right) \right\} \subseteq Z.$$

There is a good reason for the presence of these points in Z: if P denotes any of the above points, then $5[P - \infty] \in J(\mathbb{Q})$, therefore $5\lambda_\alpha(P) = 0$.

We now run through the algorithm to see that for $p = 7$ we find that

$$C(\mathbb{Q})_{\text{known}} \cup \left\{ \left((1 + \sqrt{-3})/2, \sqrt{-3} \right), \left((1 - \sqrt{-3})/2, \sqrt{-3} \right) \right\}$$

$$= Z \quad \text{up to hyperelliptic involution.}$$

First we change variables to obtain an equation for C where the defining polynomial $F(x)$ is monic, so we send $x \mapsto -4x$, $y \mapsto 4^4y$. The $\mathbb{F}_7$-points of C are

$$C(\mathbb{F}_7) = \{\overline{\infty}, \overline{(0,2)}, \overline{(0,5)}, \overline{(1,4)}, \overline{(1,3)}, \overline{(2,4)}, \overline{(2,3)}, \overline{(4,4)}, \overline{(4,3)}, \overline{(5,2)}, \overline{(5,5)}\}.$$

Of these eleven points, five of them arise as reductions of known $\mathbb{Q}$-rational points, and an order of vanishing calculation shows that these are the only rational points in those residue disks. For the remaining six $\mathbb{F}_7$-points of C, the same order of vanishing calculation shows that there is at most one $\mathbb{Q}_p$-point in each residue disk corresponding to these points on which $\lambda_{\alpha,\infty}$ and $\lambda_{\beta,\infty}$ vanish.

We know that the four quadratic points above reduce to

$$\{\overline{(1,4)}, \overline{(1,3)}, \overline{(4,4)}, \overline{(4,3)}\}$$

in some order (note that the quadratic points listed above are on the original curve, and these $\mathbb{F}_7$-points are the images after the change of variables of their reductions). Hence, our task is now reduced to the analysis of the residue disks of $\overline{(2,4)}$ and $\overline{(2,3)}$, which moreover map to each other under the hyperelliptic involution. To show that $\lambda_\alpha, \lambda_\beta$ have no zeros in these residue disks, we will explicitly write down the power series and compute their zeros using PARI/GP, as outlined in §2.1.

As usual, let $\omega_i = x^i/2y$. Then the annihilator of $J(\mathbb{Q})$ under the integration pairing is spanned by

$$\alpha = (1 + 2 \cdot 7 + 7^2 + 2 \cdot 7^3 + 5 \cdot 7^4 + O(7^5))\,\omega_0 + (4 + 7^2 + 5 \cdot 7^4 + O(7^5))\,\omega_1,$$

$$\beta = (6 + 3 \cdot 7 + 2 \cdot 7^2 + 5 \cdot 7^4 + O(7^5))\,\omega_0 + (4 + 7^2 + 5 \cdot 7^4 + O(7^6))\,\omega_2.$$

It suffices to consider the residue disk of $\overline{(2,4)}$. In this residue disk, we obtain the two power series

$$f(t) = 2 \cdot 7 + 4 \cdot 7^2 + O(7^3) + (2 + 7 + 2 \cdot 7^2 + O(7^3))t + (6 + 5 \cdot 7 + 4 \cdot 7^2 + O(7^3))t^2 + \cdots,$$

$$g(t) = 7 + 6 \cdot 7^2 + O(7^3) + (1 + 5 \cdot 7 + 2 \cdot 7^2 + O(7^3))t + (3 + 4 \cdot 7 + 7^2 + O(7^3))t^2 + \cdots.$$

Each of these have one zero in $p\mathbb{Z}_p$, but not the same zero. The two zeros are $6 \cdot 7 + 5 \cdot 7^2 + O(7^3)$ and $6 \cdot 7 + 2 \cdot 7^2 + O(7^3)$, respectively.

Example 4.3. Consider the hyperelliptic curve

$$y^2 + (x^3 + x)y = x^7 - 4x^6 + 8x^5 - 10x^4 + 8x^3 - 4x^2 + x,$$

(given above by a minimal model) which has absolute minimal discriminant 1573040 and whose Jacobian J has conductor $786520 = 2^3 \cdot 5 \cdot 7 \cdot 53^2$. We work with the odd degree model

$$C : y^2 = 4x^7 - 15x^6 + 32x^5 - 38x^4 + 32x^3 - 15x^2 + 4x,$$

on which we know the following rational points:

$$C(\mathbb{Q})_{\text{known}} = \{\infty, (0, 0), (1, -2), (1, 2)\}.$$

The point $R = [(1, -2) - \infty]$ has infinite order in $J(\mathbb{Q})$ and can thus be used to initiate the algorithm of §3.3 with $p = 11$, which is the smallest prime greater than 6 of good reduction for C. We find that

$$Z = C(\mathbb{Q})_{\text{known}} \cup W \cup \{(-1, \pm2\sqrt{-35})\},$$

where the set W of non-$\mathbb{Q}$-rational Weierstrass points has size 2. In particular, we have $C(\mathbb{Q}) = C(\mathbb{Q})_{\text{known}}$.

Perhaps more interestingly, we now explain[5] why $\{(-1, \pm2\sqrt{-35})\} \subseteq Z$. Let $K = \mathbb{Q}(\sqrt{-35})$, $\text{Gal}(K/\mathbb{Q}) = \langle \tau \rangle$ and fix an embedding $K \to \mathbb{Q}_p$. The point

$$Q = [(-1, 2\sqrt{-35}) - \infty] \in J(K)$$

is of infinite order, as there exists a non-zero differential in $H^0(J_{\mathbb{Q}_p}, \Omega^1)$ whose integral does not vanish on Q. Suppose that $nQ \in J(\mathbb{Q})$ for some integer $n \neq 0$. Then $nQ = (nQ)^\tau = nQ^\tau$ and hence $2Q = Q - Q^\tau \in J(K)_{\text{tors}}$, a contradiction. It follows that $J(\mathbb{Q})$ and Q generate a subgroup of rank 2 in $J(K)$. A computation in Magma shows that the rank of $J(K)$ itself is equal to 2. Therefore, the image of $Z \setminus W$ in $J(K)$ generates a subgroup of finite index. Showing that $(-1, \pm2\sqrt{-35}) \in Z$ is then equivalent to proving that the dimension of the p-adic closure of $J(K)$ in $J(\mathbb{Q}_p)$ is 1, i.e., that the $\mathbb{Z}$-linear independence of Q and $J(\mathbb{Q})$ is not preserved under the base change of J to $\mathbb{Q}_p$.

To explain this phenomenon, we compute the automorphism group of C using Magma. We have that $\text{Aut}(C) \cong C_2 \times C_2$, where the first copy of C_2 is generated by the hyperelliptic involution $i : C \to C$ and the second one is generated by $\varphi : C \to C$, $\varphi(x, y, z) = (z, y, x)$ which in affine coordinates corresponds to $(x, y) \mapsto (1/x, y/x^4)$. The quotient $C/\langle\varphi\rangle$ is the elliptic curve

$$E : y^2 + xy + y = x^3 - x^2,$$

whereas the quotient $C/\langle\varphi \circ i\rangle$ is the genus 2 hyperelliptic curve

$$y^2 + (x^2 + 1)y = x^5 + x^4 - 4x^3 + 3x^2 - x - 1,$$

[5]We are grateful to Andrew Sutherland for kindly computing real endomorphism algebras (using the techniques of [9, 18]) for a number of curves that produced Chabauty–Coleman output similar to this example, which greatly assisted in understanding the structure of their Jacobians. (Work by Costa et al. [13] provides further data on endomorphisms, can be made rigorous, and is freely available on GitHub.) We would also like to thank Bjorn Poonen for a very helpful discussion about this phenomenon.

which has the following odd degree model:

$$H : y^2 = 4x^5 + 5x^4 - 16x^3 + 14x^2 - 4x - 3.$$

It follows that J decomposes, over $\mathbb{Q}$, as a product of E and the Jacobian of H, which is an abelian surface A [23, Theorem 4] .

Since $\mathrm{rank}(E(\mathbb{Q})) = \mathrm{rank}(J(\mathbb{Q})) = 1$ and the p-adic closure of $E(L)$ in $J(\mathbb{Q}_p)$ can have dimension at most one for any number field L where p splits completely and any embedding $L \to \mathbb{Q}_p$, in order to determine which points of $C(\overline{\mathbb{Q}})$ can appear in Z, we then need to search for points that map to torsion points in A, under the quotient map $C \to H \to A$.

An explicit computation using Coleman integrals on H shows that $H(\mathbb{Q}_p) \cap A(\mathbb{Q}_p)_{\mathrm{tors}} \subseteq A(\mathbb{Q}_p)[2]$, where, as usual, the embedding of H into A is via the basepoint ∞. Moreover, the quotient map $C \to H$ maps ∞ and $(0, 0)$ to ∞ and is defined elsewhere by

$$(x, y) \mapsto \left(\frac{x^2 - x + 1}{x}, \frac{x^2 y - y}{x^3} \right).$$

Let $T \in C(\mathbb{Q}_p)$. Then T maps into $A(\mathbb{Q}_p)_{\mathrm{tors}}$ if and only if either $T \in \{(0, 0), \infty\}$ or $T = (x, y)$ with $y(x^2 - 1) = 0$. The latter condition is equivalent to $T \in W$ or $x^2 = 1$. This shows both why $\{(-1, \pm 2\sqrt{-35})\} \subset Z$ and why no other non-torsion point in $C(\overline{\mathbb{Q}})$ can occur in Z besides $(1, \pm 2)$ and $(-1, \pm 2\sqrt{-35})$.

4.2 Variation of the Prime and the Strong Chabauty Conjecture

We also carried out an additional set of computations in light of the following conjecture:

Conjecture 4.4 (Stoll [25, Conj. 9.5]). *Let C be a smooth projective geometrically connected curve of genus ≥ 2 over a number field K. Assume that there is a nonconstant morphism $\iota : C \to A$ into an abelian variety such that $\iota(C)$ generates A. Assume that the rank r of $A(K)$ is at most $\dim A - 2$. Then there is a finite subscheme $S \subset A$ and a set of places v of K of density 1 such that the topological closure of $A(K)$ in $A(K_v)$ meets $\iota(C)$ in a subset of $S(K_v)$.*

Remark 4.5. We thank Michael Stoll and Felipe Voloch for bringing this to our attention. As communicated to us by Stoll, the conjecture above should also have the hypothesis that the abelian variety A is simple.

With the hypothesis in the above conjecture that the abelian variety is simple, our Case 1 and 2 examples are of particular interest. We carried out the algorithm of Section 3.3 for each of these curves and all good primes between 7 and 1000 not dividing the leading coefficient of the curve's defining polynomial F. For each curve C and prime ℓ, we found that the output was just as expected, as we now explain. In what follows, the subscript ℓ (or p) added to previously defined sets emphasizes the dependence on the prime used.

Recall that for the smallest prime p satisfying the above assumptions, we had previously found that $Z_p \setminus (C(\mathbb{Q}) \cup W_p)$ was the p-adic localization of a set of points $T \subset C(K)$, for some quadratic field K. Our additional computations showed that $Z_\ell \setminus (C(\mathbb{Q}) \cup W_\ell)$ is either empty (if ℓ is inert or ramified in K) or is the ℓ-adic localization of the set T (if ℓ splits in K). In particular, this provides evidence for Conjecture 4.4 with the subscheme $S \subset J$ being the intersection of $\iota(C)$ with the saturation of $J(\mathbb{Q})$, as conjectured by Stoll.

To extend these computations to larger primes, we note that the recent work of Best [5] improving the complexity of Coleman integration algorithms would be useful for improving the speed of these computations.

References

1. J. S. Balakrishnan, *Explicit p-adic methods for elliptic and hyperelliptic curves*, Advances on superelliptic curves and their applications, NATO Sci. Peace Secur. Ser. D Inf. Commun. Secur., vol. 41, IOS, Amsterdam, 2015, pp. 260–285. MR 3525580
2. J. S. Balakrishnan, F. Bianchi, V. Cantoral-Farfán, M. Çiperiani, and A. Etropolski, *Sage code*, https://github.com/jbalakrishnan/WIN4.
3. J. S. Balakrishnan, R. W. Bradshaw, and K. S. Kedlaya, *Explicit Coleman integration for hyperelliptic curves*, Algorithmic number theory, Lecture Notes in Comput. Sci., vol. 6197, Springer, Berlin, 2010, pp. 16–31. MR 2721410
4. W. Bosma, J. Cannon, and C. Playout, *The Magma algebra system. I. The user language*, J. Symbolic Comput. **24** (1997), no. 3–4, 235–265, Computational algebra and number theory (London, 1993). MR MR1484478
5. A. J. Best, *Explicit Coleman integration in larger characteristic*, to appear, Proceedings of ANTS XIII (2018).
6. A. Booker, D. Platt, J. Sijsling, and A. Sutherland, *Genus 3 hyperelliptic curves*, http://math.mit.edu/~drew/lmfdb_genus3_hyperelliptic.txt.
7. N. Bruin and M. Stoll, *Deciding existence of rational points on curves: An experiment*, Experiment. Math. **17** (2008), no. 2, 181–189.
8. _____, *The Mordell-Weil sieve: proving non-existence of rational points on curves*, LMS J. Comput. Math. **13** (2010), 272–306. MR 2685127
9. A. R. Booker, J. Sijsling, A. V. Sutherland, J. Voight, and D. Yasaki, *A database of genus-2 curves over the rational numbers*, LMS J. Comput. Math. **19** (2016), no. suppl. A, 235–254.
10. J. S. Balakrishnan and J. Tuitman, *Magma code*, https://github.com/jtuitman/Coleman.
11. _____, *Explicit Coleman integration for curves*, Arxiv preprint (2017).
12. C. Chabauty, *Sur les points rationnels des courbes algébriques de genre supérieur à l'unité*, C.R. Acad. Sci. Paris **212** (1941), 882–885.
13. E. Costa, N. Mascot, J. Sijsling, and J. Voight, *Rigorous computation of the endomorphism ring of a Jacobian*, Mathematics of Computation (2018).
14. R. F. Coleman, *Effective Chabauty*, Duke Math. J. **52** (1985), no. 3, 765–770. MR 808103
15. _____, *Torsion points on curves and p-adic abelian integrals*, Ann. of Math. (2) **121** (1985), no. 1, 111–168. MR 782557

16. G. Faltings, *Endlichkeitssätze für abelsche Varietäten über Zahlkörpern*, Invent. Math. **73** (1983), no. 3, 349–366.
17. D. Grant, *On an analogue of the Lutz-Nagell theorem for hyperelliptic curves*, J. Number Theory **133** (2013), no. 3, 963–969. MR 2997779
18. D. Harvey and A. V. Sutherland, *Computing Hasse-Witt matrices of hyperelliptic curves in average polynomial time, II*, Frobenius distributions: Lang-Trotter and Sato-Tate conjectures, Contemp. Math., vol. 663, Amer. Math. Soc., Providence, RI, 2016, pp. 127–147.
19. N. Koblitz, *p-adic numbers, p-adic analysis, and zeta-functions*, second ed., Graduate Texts in Mathematics, vol. 58, Springer-Verlag, New York, 1984. MR 754003
20. B. Mazur, *Modular curves and the Eisenstein ideal*, Inst. Hautes Études Sci. Publ. Math. (1977), no. 47, 33–186 (1978).
21. J. S. Milne, *Jacobian varieties*, Arithmetic geometry (Storrs, Conn., 1984), Springer, New York, 1986, pp. 167–212. MR 861976
22. W. McCallum and B. Poonen, *The method of Chabauty and Coleman*, Explicit methods in number theory, Panor. Synthèses, vol. 36, Soc. Math. France, Paris, 2012, pp. 99–117.
23. J. Paulhus, *Decomposing Jacobians of curves with extra automorphisms*, Acta Arithmetica **132** (2008), no. 3, 231–244 (eng).
24. W. A. Stein et al., *Sage Mathematics Software (Version 8.1)*, The Sage Development Team, 2017, http://www.sagemath.org.
25. M. Stoll, *Finite descent obstructions and rational points on curves*, Algebra Number Theory **1** (2007), no. 4, 349–391, version of http://www.mathe2.uni-bayreuth.de/stoll/papers/covers8.pdf.
26. M. Stoll, *Arithmetic of Hyperelliptic Curves*, 2014.
27. J. L. Wetherell, *Bounding the number of rational points on certain curves of high rank*, ProQuest LLC, Ann Arbor, MI, 1997, Thesis (Ph.D.)–University of California, Berkeley. MR 2696280

Weierstrass Equations for the Elliptic Fibrations of a K3 Surface

Odile Lecacheux

Abstract We give all Weierstrass equations for the 53 elliptic fibrations of a K3 surface.

Keywords K3 surfaces · Weierstrass equations · Elliptic fibrations · Elliptic fibrations of K3 surfaces

Mathematics Subject Classification Primary 14J27, 14J28; Secondary 11G05, 06B05

1 Introduction

In previous papers [1] and [3] we classify, up to automorphisms, elliptic fibrations of the singular $K3$ surface S associated to the Laurent polynomial

$$X + \frac{1}{X} + Y + \frac{1}{Y} + Z + \frac{1}{Z} + \frac{X}{Y} + \frac{Y}{X} + \frac{Y}{Z} + \frac{Z}{Y} + \frac{Z}{X} + \frac{X}{Z}. \tag{1}$$

In the first paper, one of the fibrations

$$\pi_u : S \to \mathbb{P}^1_u$$

$$(X, Y, Z) \mapsto u = X + Y + Z$$

gives the transcendental lattice $T(S) = \langle 6 \rangle \oplus \langle 2 \rangle$ and relates S to a modular elliptic surface; the surface S can be identified with the universal elliptic curve E_u with torsion group $\mathbb{Z}/2\mathbb{Z} \times \mathbb{Z}/6\mathbb{Z}$, and $\mathbb{P}^1_u$ with the modular curve X_Γ, where

O. Lecacheux (✉)
Sorbonne Université, Institut de Mathématiques de Jussieu-Paris Rive Gauche, 4 Place Jussieu, 75252 Paris Cedex 05, France
e-mail: odile.lecacheux@imj-prg.fr

© The Author(s) and The Association for Women in Mathematics 2019
J. S. Balakrishnan et al. (eds.), *Research Directions in Number Theory*, Association for Women in Mathematics Series 19, https://doi.org/10.1007/978-3-030-19478-9_4

$\Gamma = \{ \begin{pmatrix} a & b \\ c & d \end{pmatrix} \in Sl_2(\mathbb{Z})$ with $a \equiv 1 \bmod 6, b \equiv 0 \bmod 2, c \equiv 0 \bmod 6 \}$. We complete the list of fibrations in [3], and using Kneser-Nishiyama method [9] we describe Mordell-Weil group of some of them.

This paper concludes this study by giving for each elliptic fibration a Weierstrass equation. One of the interests of such equations is the following one. The Nishiyama method which is the main subject of [1] allows to determine for each elliptic fibration the trivial lattice. If the irreducible root lattice $\mathbb{E}_6$ (resp. $\mathbb{E}_7$, $\mathbb{E}_8$, $\mathbb{D}_n$, $n \geq 4$ $\mathbb{A}_m$, $m \geq 3$) is a factor of the trivial lattice, then the fibration admits a fiber of type IV^* (resp. III^*, II^*, I^*_{n-4}, I_{m+1}). The lattices $\mathbb{A}_2$ and $\mathbb{A}_1$ can be associated to two different type of reducible fibers; to I_3 or IV for the first one; to I_2 or III for the second. We cannot distinguish between these two different cases using lattice theory. Instead, from a Weierstrass equation, Tate's algorithm allows to differentiate the various cases. All our calculations show that, except for fibration #17 (numbering used in [1]) we obtain always fibers of type I_n $n = 1, 2, 3$. The singular fibers of fibration #17 are of type I_{16}, IV, $4I_1$.

The organization of this paper is as follows: First after general recalls, we give the final results in two tables, with the elliptic parameter and x-coordinates of generators of the free part of Mordell-Weil group. Most of the fibrations are obtained from fibration π_u. For this we consider a graph G_r constructed in the paragraph 4. We can compare this graph to the famous Coxeter graph corresponding to the universal elliptic curve with 7-torsion [5] or to the graph of [2] Figure 3. Then we list the used methods and give only details of Weierstrass equations for fibrations #1, #51, and #8. The remaining fibrations need to add some rational curves. They are constructed as components of singular fibers of previous fibrations. For instance, we start from fibration #8 as explained in paragraph 7, and use a component of a I_2 fiber. Others fibrations are obtained from fibrations #33, #26, with sections of infinite order. The last fibrations are characterized by some other features. The paragraph 11 deals with the following problem; from Weierstrass equations we have to distinguish two fibrations resulting from different embeddings in Niemeier lattices with the same trivial lattice.

Notation 1. *For every elliptic fibration, the final result is a Weierstrass equation with x and y for the variables, the parameter is always denoted by t in Tables 1 and 2. The singular fibers of type R at $t = a, b..$ or at a root of $p(t)$ are denoted $R(a, b..)$ or $R(p(t))$.*

Computations were performed using the computer algebra system MAPLE and the Maple Library "Elliptic Surface Calculator" written by Kuwata [8].

2　General Results on Elliptic $K3$ Surfaces

Let k be an algebraically closed field. In this paper, an elliptic $K3$ surface S is a $K3$ surface with genus 1 fibration $\pi : S \to \mathbb{P}^1$, and a section O. So, we have a Weierstrass equation

$$y^2 + a_1(t)xy + a_3(t)y = x^3 + a_2(t)x^2 + a_4(t)x + a_6(t)$$

with $a_i(t)$ of degree at most $2i$, and at least one of degree $2i$. We call t the elliptic parameter. This Weierstrass equation describes the generic fibers F.

The singular fibers, classified by Néron and Kodaira, are in the following Kodaira types:

- two infinite series $I_n(n \geq 1)$ and $I_n^*(n \geq 0)$,
- five types $II, III, IV, II^*, III^*, IV^*$.

The reducible fibers are union of components with multiplicities; each component is a smooth rational curve with self-intersection -2. The non-identity components of any reducible fiber $\pi^{-1}(v)$ contribute an irreducible root lattice (scaled by (-1)), T_v to the Néron-Severi lattice $NS(S)$. The trivial lattice is defined by

$$T = \langle O, F \rangle \oplus (\oplus_v T_v).$$

Notice that O and F span a copy of the hyperbolic plane U.

The dual graph of these components (a vertex for each component, an edge for each intersection point of two components) is an extended Dynkin diagram of type $\tilde{\mathbb{A}}_n, \tilde{\mathbb{D}}_l, \tilde{\mathbb{E}}_p$. Deleting the zero component (i.e., the component meeting the zero section) gives the Dynkin diagram graph $\mathbb{A}_n, \mathbb{D}_l, \mathbb{E}_p$.

We draw the most useful diagrams, with the multiplicity of the components, the zero component being represented by a circle (Figure 1).

A theorem of Shioda and Tate [13, 14] shows that as long as the elliptic fibration is non-trivial (equivalently, it has at least one singular fiber), the Mordell-Weil group of F over $\mathbb{P}^1$, $MW(S/\mathbb{P}^1)$ is isomorphic to $NS(S)/T$.

In particular, we have the Shioda-Tate formulas

$$\rho(S) = 2 + \text{rank } MW(S/\mathbb{P}^1) + \sum_v \text{rank } T_v$$

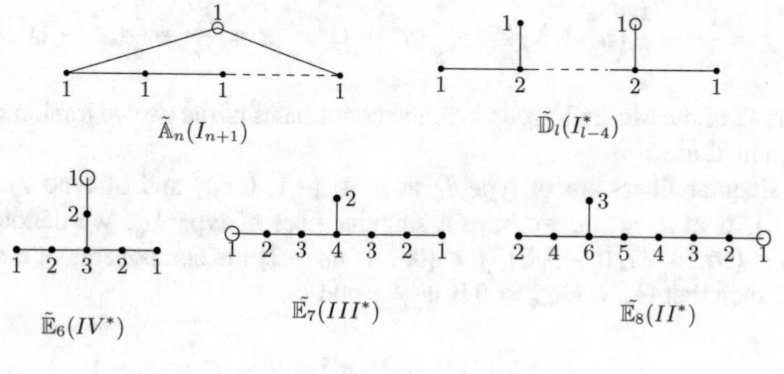

Fig. 1 Extended Dynkin diagrams

and

$$|\text{discr}\,(NS\,(S))| = \frac{\det(\Delta)\,\Pi_v \text{discr}\,(T_v)}{|\,MW_{tor}\,|^2},$$

where $\rho(S)$ is the rank of $NS(S)$, Δ the height pairing matrix for a basis of the torsion-free part of the Mordell-Weil group of F over $\mathbb{P}^1$. Heights are computed with Weierstrass equation as explained in [15]. Using the previous formula we can obtain generators of the Mordell-Weil group.

3 The Results

See Table 1 and 4.

4 The Modular Fibration and the Graph Gr

The modular fibration is obtained with the elliptic parameter $u = X+Y+Z$ from (1) and the following transformations:

$$X = -\frac{y + x(u+1)}{x\,(u-3)}, \qquad Y = -\frac{y(u+1) - x^2}{y\,(u-3)},$$

$$Z = u - X - Y = -\frac{(u+1)\,(x + (u+1)\,(u-1)^2)}{x\,(u-3)}$$

$$E_u : y^2 + \left(u^2 + 3\right) xy + \left(u^2 - 1\right)^2 y = x^3. \tag{2}$$

We use also the following equation and transformations:

$$Y'^2 = X' \left(X' - (u-3)\,(u+1)^3\right) \left(X' - (u+3)\,(u-1)^3\right) \tag{3}$$

$$y = \frac{Y'}{8} - \frac{1}{8}\left(u^2 + 3\right) X' + \frac{1}{8}(u^2 - 1)^3, \qquad x = \frac{X'}{4} - \frac{1}{4}(u^2 - 1)^2.$$

The rank of the Mordell-group is 0, the coordinates of the twelve torsion sections are given in Table 3.

The singular fibers are of type I_6 at $u \in \{-1, 1, \infty\}$ and of type I_2 at $u \in \{3, -3, 0\}$. If at $u = u_0$, we have a singular fiber of type I_{n_0}, we denote $\Theta_{u_0, j}$ with $u_0 \in \{\infty, -1, 1, 0, -3, 3\}$, $j \in \{0, \ldots, n_0 - 1\}$ the components of a singular fiber I_{n_0} such that $\Theta_{w,j} \cdot \Theta_{v,k} = 0$ if $w \neq v$ and

$$\Theta_{u_0,j} \cdot \Theta_{u_0,k} = \begin{cases} 1 \text{ if } |k - j| = 1 \text{ or } |k - j| = n_0 - 1 \\ -2 \text{ if } k = j \\ 0 \text{ otherwise,} \end{cases}$$

Table 1 Results 1

#no	Equation	From	x-Coordinate
#1	$y^2 = x^3 - 15\left(t^2 - 1\right)^2 x^2 - 16\left(t^2 - 1\right)^5$	#50	$\frac{4}{81} + \frac{5}{81}\left(t^2 - 1\right)\left(125t^4 - 118t^2 + 17\right)$
#2	$y^2 = x^3 - 3t^3\left(5t - 4\right)x - 2t^5\left(2t^2 - 5t + 20\right)$	#8	
#3	$y^2 = x^3 + \left(t^3 - 15t^2 + 8\right)x^2 + 16x$	#1	$\frac{1}{144}\frac{(t-8)^2\left(t^2-16t+40\right)^2}{(t+10)^2}$
#4	$y^2 = x^3 + 2t^2\left(2t + 3\right)x^2 + 17t^4x + 4t^5$	#8	$1 - 4t$
#5	$y^2 = x^3 + \left(t^3 - 15t + 10\right)x^2 + (-48t + 160)x$	#26	
#6	$y^2 = x^3 + t^2\left(t - 1\right)\left(4t^2 + 9t + 3\right)x - \frac{2}{3}t^3\left(2t^4 - 15t^3 - 3t^2 + 9t + 3\right)$	#19	$\frac{1}{9}t\left(4t^3 + 12t^2 - 9t - 9\right)$
#7	$y^2 = x^3 - t\left(15t - 4\right)x^2 + t^5x$	#3	$\frac{1}{144}\frac{\left(t^3-24t^2+48t-16\right)^2}{(t+2)^2}$
#8	$y^2 = x^3 + 6t^2x^2 - t^3\left(4t - 1\right)\left(t - 4\right)x$	#50	$1 - 4t$
#9	$y^2 = x^3 + t\left(t^2 + 9t + 12\right)x^2 - 4t^3x$	#11	4
#10	$y^2 = x^3 + t\left(t^2 - 6t - 3\right)x^2 + 16t^3x$	#12	
#11	$y^2 = x^3 + t^2(t + 9)x^2 + 4t^3(3t - 1)x$	#8	$-t^3$
#12	$y^2 = x^3 + 2t^2\left(2t - 3\right)x^2 - t^3\left(3t - 4\right)x$	#50	
#13	$y^2 = x^3 - 12t\left(t - 1\right)x^2 - 64t^3\left(t - 1\right)^2 x$	#33	$-4(t - 1)$
#14	$y^2 - \left(15 + t^2\right)xy - 4y = x^3$	#26	$\frac{1}{324}\left(t^6 + 9t^4 + 27t^2 - 81\right)$
#15	$y^2 = x^3 + (t + 1)\left(3t + 1\right)x^2 - 2t^3\left(t^2 - 4t - 2\right)x + 4t^6$	#33	$0; -6t$
#16	$y^2 + (t - 3)\left(t + 3\right)xy - 4y = x^3 - 12x^2$	#9	$0; -2(t - 3)$
#17	$y^2 - \left(t^2 + 6\right)xy - 16y = x^3 + 3x^2$	#33	0
#18	$y^2 + \left(t^2 + 2t + 4\right)xy - 4t\left(t + 2\right)y = (x - 4)\left(x - 2t\right)^2$	#27	$2t$
#19	$y^2 = x^3 + \frac{1}{4}\left(t - 3\right)\left(t^2 + 8\right)x^2 + \left(-2t^2 + 1\right)x + 12 + 4t$	#4	$4\frac{(t-1)}{(t-2)^2}$
#20	$y^2 = x^3 - t\left(t^2 - 6t + 17\right)x^2 + 16t^2x$	#8	$t(t - 1)^2$
#21	$y^2 = x^3 - t\left(2t^2 + 3t + 4\right)x^2 + 2t^2\left(3t + 2\right)x$	#33	$2t(t + 1)^2$
#22	$y^2 = x^3 + 2t\left(2t^2 + 3t + 2\right)x^2 + t^4x$	#50	1
#22 bis	$y^2 = x^3 - t\left(t^2 - 6t + 1\right)x^2 + 16t^4x$	#50	$t(t - 1)^2$
#23	$y^2 = x(x + 4t)(x + t^2(t - 3))$	#8	4
#24	$y^2 + \left(6 - t^2\right)xy - x\left(x - 16\right)\left(x - 1\right)$	#8	$1; (t - 2)^2$
#25	$y^2 - 2\left(t + 4\right)xy - 4t^2y = x\left(x - 12\right)\left(x + 2t^2 + t^3\right)$	#34	$0; 12$
#26	$y^2 + \left(4t^2 + 3\right)xy - 12y = x(x - 4)^2$	#23	$-4\left(2t - 1\right)\left(t + 1\right); -4\left(2t + 1\right)\left(t - 1\right)$

(continued)

Table 1 (continued)

#no	Equation	From	x-Coordinate
#27	$y^2 - t(t-2)xy + 4t^2 y = x(x - 2t(t-1))^2$	#48	$2t(t-1)$
#28	$y^2 + 6txy - 2t^2(t-1)^2 y = x^3$	#50	
#29	$y^2 - 4txy - 4t^3(t+1)y = x^2(x - t(t+1))$	#50	$0; (2t+1)(t+1)$
#30	$y^2 = x^3 + t(t+2)(t^2+4t+1)x^2 + t^2 x$	#50	
#31	$y^2 + (1-6t)xy - 2t^3 y = x^2(x - 2t^3)$	#16	$2t^3$
#32	$y^2 + (t^2+3)xy - 4t^2 y = x^3$	#50	$t^2; 2(t-1)$
#33	$y^2 = x^3 + t^2(t^2-3)x^2 - 2t^4 x + t^4$	#50	$0; 2t+1; -2t+1$
#34	$y^2 - 2txy - 2t^2(t+3)(t+1)y = x^2(x+2t(t+1))$	#50	0
#35	$y^2 + (2t-3)xy + 2t^2(t-1)y = x^2(x-2t^2(t+1))$	#50	0
#36	$y^2 = x(x+t^2(4t+1))(x-t(4t+1))$	#50	$-4t^3$
#37	$y^2 = x^3 + t(4t^3-8t^2+3t-1)x^2 + t^3 x$	#50	$t; 4t^3$
#38	$y^2 - t(t+2)xy - 2t^3(t+1)^2 y = x(x+2t^2(t+1))(x+2t(t^2-1))$	#26	$0; -2t^2(t+1)$
#39	$y^2 = x^3 + t(t^3-2t^2+3t-4)x^2 - t^2(2t-3)x$	#50	t

the dot meaning the intersection product. By definition, the component $\Theta_{u_0,0}$ intersects the zero section (0). The n_0-gon obtained can be oriented in two ways for $n_0 > 2$. For each u_0 we want to know which component is cut off by a torsion section (M), i.e., the index $r(M, u_0)$ such that $M.\Theta_{u_0,r(A,u_0)} = 1$. We choose the orientation for $r(P_3, u_0)$ to be $< \frac{n_0}{2}$. For example, if $M = P_3$ or A_2 and $u_0 = 1$, denote $T = u - 1$ and use Equation (3)

$$Y'^2 = X'^3 + \left(16 + 16T + 8T^2 + 2T^4\right)X'^2 + T^3(T-2)(T+4)(T+2)^3 X',$$

then $P_3 = \left(X'_{P_3} = T^2(T+2)^2, Y'_{P_3} = 4T^2(T+2)^2\right)$. The components $\Theta_{1,1}$ and $\Theta_{1,5}$ are obtained after one blow up $X' = TX'_1, Y' = TY'_1$ and are defined by $Y_1'^2 = 16X_1'^2$. The components $\Theta_{1,2}$ and $\Theta_{1,4}$ are obtained after a second blow up $X' = T^2X'_2, Y' = T^2Y'_2$. Since we have $Y'_{P_3} + 4X'_{P_3} = 8T^2(T+2)^2$ and $Y'_{P_3} - 4X'_{P_3} = 0$, then $r(P_3, u_0) = 2$ or 4. We choose the orientation for $r(P_3, 1) = 1$, then $\Theta_{1,2}$ correspond to $Y'_2 - 4X'_2 = 0$. The computation $r(A_{23}, u_0)$ and $r(A_{22}, u_0)$ can be done in the same way and by addition for the other sections. We obtain Table 4. In this table we can see that each component of reducible fibers of type I_6 is cut off by exactly two sections.

This table already appears in ([6] Table 2), computed in a theoretical way.

From this table we can draw the graph Gr. The vertices are the 12-torsion sections and the 24 components of singular fibers, two vertices are linked by an

Table 2 Results 2

#no	Equation	From	x-Coordinate
#40	$y^2 + 2\left(t^2 - 1\right)xy - 2t^2y = x\left(x + 4t^2\right)\left(x + t^2\right)$	#50	$0; -2t^2(t+1)$
#40-bis	$y^2 - \left(t^2 + 1\right)xy + 4t^2y = x\left(x - t^2\right)\left(x - 4t^2\right)$	#50	$0; 2t(t-1)$
#41	$y^2 = x^3 - 2t\left(t^2 + 1\right)x^2 + t^2\left(t+1\right)^4 x$	#50	$t^2(t+1)^2$
#42	$y^2 - 2t^2xy = \left(x + (t-1)^2\right)\left(x - (t-1)^3(t+1)\right)\left(x + 1 - t^2\right)$	#50	$t^2 - 1; -(t-1)^2$
#43	$y^2 - t(t-2)xy = x(x - 2t(t+1))^2$	#50	$2t$
#44	$y^2 + \left(t^2 - 4t + 1\right)xy - 4t^3y = x^3$	#50	$4t; -2t(t-1)$
#45	$y^2 + \left(-2t^2 + 2t + 1\right)xy - 2t^3y = x\left(x + 4t^3\right)(x + t)$	#50	$0; -4t^3$
#46	$y^2 + \left(-\frac{1}{4}t^2 + \frac{5}{3}\right)xy + (2t - 8)y = x(x - 6 - 3t)\left(x - \frac{1}{6}t^2 - \frac{4}{9} + t\right)$	#26	$0; 3t + 6$
#47	$y^2 + \left(-t^2 + 4t - 5\right)xy - 4ty = x(x + 2t - 6)\left(x - 2t^2 + 2t\right)$	#26	$0; 6 - 2t$
#48	$y^2 + \left(t^2 - 2\right)xy + 2t^2(t - 2)y = x\left(x - 2t + 5t^2\right)\left(x + 2t^2\right)$	#26	$0; -2t^2$
#49	$y^2 = x^3 + 1/4\left(t^2 - 10t + 1\right)(t+1)^2 x^2 + 4t^3(t+1)^2 x$	#50	$2t(t+1)$
#50	$y^2 + \left(t^2 + 3\right)xy + (t - 1)^2(t+1)^2 y = x^3$		
#51	$y^2 + \left(2t^2 + 1\right)xy + 2t^2(t - 1)y = x\left(x + t^2 - t\right)\left(x - 2t^2\right)$	#50	$0; 2t^2$

Table 3 Coordinates of torsion sections.

Sections	$A_2 : (2A_2 = 0)$	$A_{22} : (2A_{22} = 0)$	$A_{23} : (2A_{23} = 0)$	P_3	$2P_3 : (3P_3 = 0)$
$x:$	$-\frac{1}{4}\left(u^2 - 1\right)^2$	$-(u+1)^2$	$-(u-1)^2$	0	0
$y:$	$\frac{1}{8}\left(u^2 - 1\right)^3$	$(u+1)^3$	$-(u-1)^3$	0	$-\left(u^2 - 1\right)^2$
Sections	$A_2 + P_3$	$A_{22} + P_3$	$A_{23} + P_3$		
$x:$	$-2\left(u^2 - 1\right)$	$-(u-1)^2(u+1)$	$(u+1)^2(u-1)$		
$y:$	$8\left(u^2 - 1\right)$	$(u-1)^4(u+1)$	$-(u+1)^4(u-1)$		
Sections	$A_2 + 2P_3$	$A_{22} + 2P_3$	$A_{23} + 2P_3$		
$x:$	$-2\left(u^2 - 1\right)$	$-(u-1)^2(u+1)$	$(u+1)^2(u-1)$		
$y:$	$\left(u^2 - 1\right)^2$	$\left(u^2 - 1\right)^2$	$\left(u^2 - 1\right)^2$		

edge if they intersect. To have a readable picture we divide into two graphs *Graph I* and *Graph II* (Figures 2 and 3). Only the generators of the torsion sections are drawn on *Graph II*.

Table 4 Intersections

	P_3	$2P_3$	A_2	A_{22}	A_{23}
$u = 1$	$r(P_3, 1) = 2$	$r(2P_3, 1) = 4$	3	0	3
$u = -1$	$r(P_3, -1) = 2$	$r(2P_3, -1) = 4$	3	3	0
$u = \infty$	$r(P_3, \infty) = 2$	$r(2P_3, \infty) = 4$	0	3	3
$u = 3$	0	0	1	1	0
$u = -3$	0	0	1	0	1
$u = 0$	0	0	0	1	1

$A_2 + P_3$	$A_{22} + P_3$	$A_{23} + P_3$	$A_2 + 2P_3$	$A_{22} + 2P_3$	$A_{23} + 2P_3$
5	2	5	1	4	1
5	5	2	1	1	4
2	5	5	4	1	1
1	1	0	1	1	0
1	0	1	1	0	1
0	1	1	0	1	1

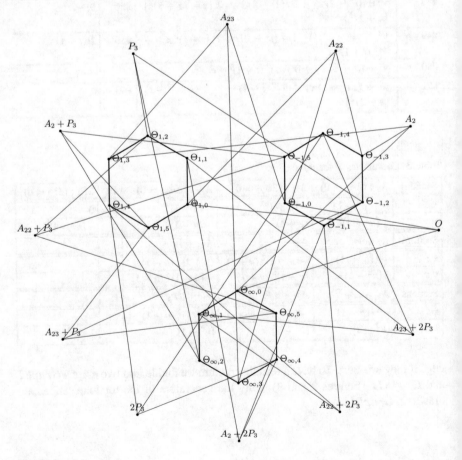

Fig. 2 Graph I

Fig. 3 Graph II

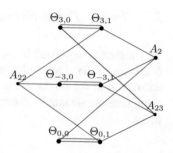

5 Methods

First we will use the following proposition ([10] pp. 559–560 or [12] Prop. 12.10).

Proposition 2. *Let S be a K3 surface and D an effective divisor on S that has the same type as a singular fiber of an elliptic fibration. Then S admits a unique elliptic fibration with D as a singular fiber. Moreover, any irreducible curve C on S with $D.C = 1$ induces a section of the elliptic fibration.*

If S is a $K3$ surface, an elliptic fibration $f : S \to \mathbb{P}^1$ with a section 0 defines a nonconstant function t, with $t = f(z)$ for $z \in S$; the function t, called the elliptic parameter is unique (up to a homographic transformation). Then the generic fiber F has a Weierstrass equation on $k(t)$. From previous proposition, to construct a fibration we need one effective divisor D, we call such a divisor an elliptic divisor.

5.1 Graph Method

Let S be a $K3$ surface and $f : X \to \mathbb{P}^1$ an elliptic fibration of elliptic parameter t. Let Δ be a set of components of reducible fibers included in $f^{-1}(W)$, where W is a finite subset of $\mathbb{P}^1$ and δ a finite set of sections. Suppose D_1 and D_2 are two elliptic divisors of the same new fibration with supports contained in $\Delta \cup \delta$. If we write $D_i = \delta_i + \Delta_i$ with δ_i sum of sections and Δ_i sum of components of reducible fibers, then $\delta_1 - \delta_2$ is the divisor of a function u_0 on the generic fiber F and a parameter of the new fibration can be chosen as the function $u = u_0 \prod_{t_i \in W} (t - t_i)^{a_i}$, where $a_i \in \mathbb{Z}$.

Indeed, for the new fibration, there is a function u with divisor $D_1 - D_2$ so D_1 and D_2 belong to the same class in $NS(S)$. As the class of $\Delta_1 - \Delta_2$ is in T, then δ_1 and δ_2 are in the same class in $NS(S)/T$. Using the isomorphism between $NS(S)/T$ and $F(k(t))$, it follows that $\delta_1 - \delta_2$ is the divisor of a function u_0 on $F(k(t))$.

All the elements of $\Delta \cup \delta$ are the vertices of a graph G_r. Two vertices e, f are linked by an edge if $e.f = 1$. Then following the extended Dynkin diagrams representing reducible fibers we look for subgraph representing D_1 and D_2.

5.2   Two or Three-Neighbor Method

The method is explained in [4, 7, 17] or [11] §6. The components of the reducible fibers are obtained after blow up at singular points, so it is convenient to make a translation of the origin of coordinates to have all singular points in $(0, 0)$. Then using Chinese remainder theorem we can have this property for several reducible fibers.

5.3   To Weierstrass Form

Once we get an elliptic parameter t, we have to compute a Weierstrass equation by birational transformations. We can eliminate one variable in function of the new parameter and most time we obtain an equation of bidegree 2 in the others variables. Blowing up the singular points we obtain then a Weierstrass equation.

If we have a quartic equation $y^2 = ax^4 + bx^3 + cx^2 + e^2$, with $e \neq 0$, setting $y = e + \frac{dx}{2e} + x^2 X'$, $x = \frac{8c^3 X' - 4ce^2 + d^2}{Y'}$ we get

$$Y'^2 + 4e \left(dX' - be \right) Y + 4e^2 \left(8e^3 X' - 4ce^2 + d^2 \right) \left(X'^2 - a \right) = 0.$$

Finally the Weierstrass equation follows easily.

If we have a cubic equation with a rational point we can use [16] page 23 or a software program (Maple, Sage).

All these transformations will be called standard.

6   Fibrations from the Graph Gr and the Modular Fibration

In this section, we derive as much as possible of fibrations from the fibration #50 using the methods previously explained. The results are given in Table 5. Only some cases are explained.

6.1   Fibration #1

Let D and D' be the two divisors

$$D = 6\,(0) + 4\Theta_{1,0} + 2\Theta_{1,5} + 3\Theta_{-1,0} + 5\Theta_{\infty,0} + 4\Theta_{\infty,5} + 3\Theta_{\infty,4} + 2\,(2P_3 + A_2) + \Theta_{3,1}$$

$$D' = 6\,(P_3) + 4\Theta_{1,2} + 2\Theta_{1,3} + 3\Theta_{\infty,2} + 5\Theta_{-1,2} + 4\Theta_{-1,3} + 3\Theta_{-1,4} + 2\,(2P_3 + A_{23}) + \Theta_{0,1}.$$

Table 5 Two elliptic divisors in Gr

Fibration	Parameter
#1 − h $II^*(\infty, 0)$, $I_2(-2)$, $2I_1$	$\dfrac{y^2\left(y-2(u-1)x+(u^2-1)^2\right)}{(u-1)^4(u+1)^3\left(y+(u^2-1)x+(u^2-1)^2\right)}$
#8 − c $III^*(\infty, 0)$, $I_2\left(1, 4, \frac{1}{4}\right)$	$\dfrac{y}{(u-1)^4}$
#12 − l $I_4^*(\infty)$, $III^*(0)$, $I_3(1)$, $I_2\left(\frac{4}{3}\right)$	$\dfrac{y+4x}{(u-1)^2}$
#22 − n $I_4^*(\infty)$, $I_4^*(0)$, $I_2(-1)$, $2I_1$	$\dfrac{y+x(u-1)^2}{(u-1)^2\left(x+(u-1)^2\right)}$
#22 − n′ $I_4^*(\infty)$, $I_4^*(0)$, $I_2(-1)$, $2I_1$	$\dfrac{(u-1)^2\left(y+x(u-1)^2\right)}{y\left(x+(u-1)^2\right)}$
#28 − b $IV^*(\infty, 0)$, $I_6(1)$, $I_2(-1)$	$\dfrac{y}{(u^2-1)^2}$
#29 − δ $IV^*(\infty)$, $I_3^*(0)$, $I_4(-1)$, $3I_1$	$\dfrac{XY(1-X+Y+Z)(X+Y+Z-1)}{Z(YZ+X)(X+Y+Z+1)^2} = \dfrac{y(u+1)-x^2}{(u+1)^3\left(x+(u-1)^2(u+1)\right)}$
#30 − e $I_{12}(\infty)$, $I_1^*(0)$, $I_2(-3; -1)$, $I_1(-4)$	$\dfrac{4x+(u^2-1)^2}{x(u^2-1)}$
#32 − s $I_{12}(\infty)$, $I_6(0)$, $6I_1$	$\dfrac{2y+(u-3)(u-1)x-(u-1)^3(u+1)^2}{x(u^2-1)}$
#33 − β $I_{10}(\infty)$, $IV^*(0)$, $6I_1$	$\dfrac{-(u-1)^3\left(y+u(u-1)x+(u+1)(u-1)^3\right)}{2\left(y+x(u-1)^2\right)\left(x+(u-1)^2\right)}$
#34 − m $IV^*(\infty)$, $I_1^*(0)$, $I_6(-1)$, $I_2(-3)$, $I_1\left(\frac{-32}{27}\right)$	$\dfrac{y+4x}{(u^2-1)^2}$
#35 − θ $I_3^*(\infty)$, $I_6(0, 1)$, $3I_1$	$\dfrac{-\left(x+(u+1)(u-1)^2\right)\left(y+(u^2+3)x+2(u^2-1)^2\right)}{(u+1)(u-1)^2\left(y+x(u-1)^2\right)}$
#36 − w $I_2^*(\infty, 0)$, $I_2(-1)$, $I_0^*\left(-\frac{1}{4}\right)$	$\dfrac{x}{(u^2-1)^2}$
#37 − ε $I_{10}(\infty)$, $I_2^*(0)$, $I_2(1)$, $4I_1$	$\dfrac{2y+x(u^2-1)}{\left(4x+(u^2-1)^2\right)(u^2-1)}$
#39 − v $I_{10}(\infty)$, $I_0^*(0)$, $I_4(1)$, $I_2\left(\frac{3}{2}\right)$, $2I_1$	$\dfrac{(Y+Z)(X+Y+Z-3)}{(X+Y+Z)^2-1} = \dfrac{y+x(u-1)^2}{x(u^2-1)}$
#40bis − t $I_{10}(\infty)$, $I_8(0)$, $6I_1$	$\dfrac{1}{2}\dfrac{2y+(u-3)(u-1)x-(u-1)^3(u+1)^2}{(u^2-1)\left(x+\frac{1}{4}(u^2-1)^2\right)}$
#40 − p $I_{10}(\infty)$, $I_8(0)$, $6I_1$	$\dfrac{\left(x-(u-1)(u+1)^2\right)\left(y(1-u^2)+2x^2\right)}{(u^2-1)\left(y+(u^2+3)x+2(u^2-1)^2\right)}$
#41 − r $I_1^*(\infty, 0)$, $I_8(-1)$, $2I_1$	$\dfrac{XY}{Z} = \dfrac{x+(u+1)^2}{(u+1)(u-3)}$
#42 − ζ $I_1^*(\infty, 0)$, $I_6(1)$, $I_2(-1)$, $2I_1-$	$\dfrac{Y+1}{X+1} = -\dfrac{y+(u^2-1)x+(u^2-1)^2}{y+4x}$

(continued)

Table 5 (continued)

Fibration	Parameter
#43 − k $I_8\,(\infty)$, $I_1^*\,(0)$, $I_4\,(-1)$, $I_2\,(-2,2)$, $I_1\,(8)$	$X + Y = \dfrac{(u-1)^2\left(x+(u+1)^2\right)}{(u-3)x}$
#44 − t $I_9\,(\infty, 0)$, $6I_1$	$-\dfrac{y-8(u^2-1)}{(u-1)(x+2(u^2-1))} + \dfrac{y-(u^2-1)^2}{(u+1)(x+(u+1)(u-1)^2)} + \dfrac{u-5}{u-1}$ $= \dfrac{(y(1-u^2)+2x^2)(x+(u+1)(u-1)^2)}{(u^2-1)(x+2(u^2-1))(y+x(u-1)^2)}$
#45 − j $I_9\,(\infty)$, $I_7\,(0)$, $I_3\left(\frac{-1}{2}\right)$, $5I_1$	$\dfrac{y+(u^2+3)x+2(u^2-1)^2}{(u^2-1)(x+(u+1)(u-1)^2)}$
#49 − z $I_6\,(\infty, 0)$, $I_0^*\,(-1)$, $I_4\,(1)$, $2I_1$	$Z = -\dfrac{(u+1)(x+(u+1)(u-1)^2)}{x(u-3)}$
#51 − f $I_7\,(\infty, 0)$, $I_5\,(-1)$, $5I_1$	$\dfrac{2y+x(u^2-1)}{(u-1)\left(4x+(u^2-1)^2\right)}$

We have $D.D' = 0$ and D and D' are of type II^*. They are two singular fibers of an elliptic fibration. The horizontal divisor of $D - D'$ is equal to $6P_3 - 6\,(0) + 2\,(2P_3 + A_{23}) - 2\,(2P_3 + A_2)$, it is the divisor of the function $\dfrac{y^2\left(y-2(u-1)x+(u^2-1)^2\right)}{y+(u^2-1)x+(u^2-1)^2}$ on E_u. We can choose h as an elliptic parameter with

$$h = \frac{1}{(u-1)^4\,(u+1)^3}\,\frac{y^2\left(y-2(u-1)x+(u^2-1)^2\right)}{y+(u^2-1)x+(u^2-1)^2}.$$

To obtain a Weierstrass equation we eliminate x between the equation of E_u and h. After resolving the singularities we obtain the smooth cubic equation

$$U^3 + 2hV^3 + 4hV^2 - 6hUV - 6hU - h(h-2)V = 0.$$

Then after standard transformation we deduct the following Weierstrass equation:

$$Y'^2 + \frac{1}{3}\frac{(h-2)\left(h^2+104h+4\right)}{(h+2)}X'Y' + 648h^2\,(h-2)\,(h+2)\,Y'$$
$$= \left(X' - \frac{h^6 - 444h^5 - 1668h^4 - 83104h^3 - 6672h^2 - 7104h + 64}{36\,(h+2)^2}\right)$$
$$\left(X'^2 - 7776h^3\,(h+2)^2\right).$$

After simplification a short Weierstrass form is obtained

$$y^2 = x^3 - 75xh^4 + 2h^5(h^2 - 121h + 4) \tag{4}$$

with the rational point

$$P_h : \begin{pmatrix} x = \frac{1}{1296} \frac{(h^6-228h^5+2220h^4-69280h^3+8880h^2-3648h+64)}{(h+2)^2}, \\ \\ y = \frac{1}{46656} \frac{(h-2)(h^8-340h^7+22144h^6+301520h^5-6976544h^4+1206080h^3+354304h^2-21760h+256)}{(h+2)^3} \end{pmatrix}.$$

The singular fibers are II^* $(0, \infty)$, I_2 (-2), I_1 $(h^2 - 246h + 4)$. The Mordell-Weil group is of rank one generated by P_h.

We have also the equation with $s = \frac{-(h-2)}{(h+2)}$

$$Y^2 = X^3 - 15(s^2 - 1)^2 X^2 - 16(s^2 - 1)^5$$

$$P_s = (\frac{4}{81} + \frac{5}{81} (s-1)(s+1)\left(125 s^4 - 118 s^2 + 17\right),$$

$$-s\left(10935 - 51030 s^2 + 85212 s^4 - 60750 s^6 + 15625 s^8\right)).$$

6.2 Fibration #51

We consider the divisors D, D', D'' of type, respectively, I_7, I_7, I_5

$$D = (0) + \Theta_{\infty,0} + (A_2) + \Theta_{1,3} + \Theta_{1,4} + \Theta_{1,5} + \Theta_{1,0}$$

$$D' = (P_3) + \Theta_{-1,2} + \Theta_{-1,1} + (A_2 + 2P_3) + \Theta_{\infty,4} + \Theta_{\infty,3} + \Theta_{\infty,2}$$

$$D'' = (A_{23} + 2P_3) + \Theta_{-1,4} + \Theta_{-1,5} + (A_{22} + P_3) + \Theta_{0,1}.$$

Since we have $D.D' = D.D'' = 0$, they correspond to 3 singular fibers of the same elliptic fibration. So there exists a unique elliptic parameter for this fibration, m such that $m(D') = 0$, $m(D) = \infty$, and $m(D'') = 1$. The horizontal divisor (compared to the modular fibration #50) of $D' - D$ is equal to $P_3 + (A_2 + 2P_3) - 0 - A_2$, so $m = k\frac{2y+(u^2-1)x}{4x+(u^2-1)^2}$; moreover, the condition $m(D'') = m(A_{22} + P_3) = 1$ gives $k = \frac{1}{u-1}$ and the Weierstrass equation

$$y^2 + \left(2m^2 - 1\right) yx + 2m^2 (m-1)(2m-1) y$$

$$= (x + m(m-1))\left(x + 2m^2(m-1)\right)\left(x - 2m^2(m-1)\right).$$

The singular fibers are I_7 $(0, \infty)$, I_5 (1), I_1 $\left(8m^5+16m^4+20m^3-52m^2+6m-1\right)$.

The Mordell-Weil group is of rank two, generated by the two points $(0, -2m^3 (m-1))$ and $(-m(m-1), 0)$ of infinite order.

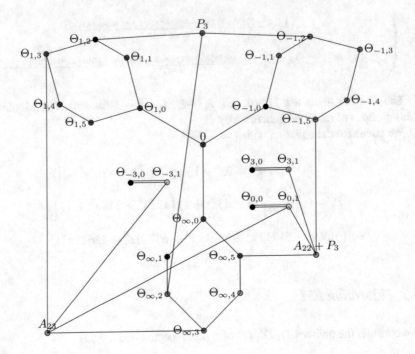

Fig. 4 Fibration #8 part of *Gr*

6.3 Fibration #8

Figure 4 is a part of *Gr* showing elliptic divisors for fibration #8.

We consider the two divisors: the divisor D^- (in blue in Figure 4) of type III^*

$$D^- = \Theta_{1,3} + 2\Theta_{1,4} + 3\Theta_{1,5} + 4\Theta_{1,0} + 2\Theta_{1,1} + 3(0) + 2\Theta_{\infty,0} + \Theta_{\infty,5}$$

and (in red on the graph) the divisor D' of type III^*

$$D^+ = \Theta_{\infty,3} + 2\Theta_{\infty,2} + 3(P_3) + 4\Theta_{-1,2} + 3\Theta_{-1,3} + 2\Theta_{-1,4} + \Theta_{1,5} + 2\Theta_{-1,1}.$$

In pink, yellow, and orange we have also parts of other singular fibers of the same fibration. In green we have three sections of the fibration.

For the elliptic parameter, with the previous method we can choose $c = \frac{y}{(u-1)^4}$. At $u = 0$ (resp. $u = 3$ $u = -3$), the singularities are for $y = 1$ (resp. $y = 64$, $y = 64$), so the value of c on $\Theta_{0,1}$ is 1, on $\Theta_{3,1}$, 4, and on $\Theta_{-3,1}$, 1/4.

We substitute $y = c(u-1)^4$ in the equation of E_u and define w by $x = w(u-1)^2$. So we have the equation of a cubic curve with the rational point $A_{23} : w = 1, u = \frac{t-1}{t}$. We translate the origin of coordinates in this point and change axes to have the line $Z = 0$ as the tangent in this point. Then we obtain a cubic equation of the shape

$$F(U, Z)Z + U^2 R(U, Z) = 0.$$

The standard transformations give a Weierstrass equation with the point A_{23} as the 0 of the group law on the new fibration. Precisely with the following transformations:

$$u = \frac{(c-1)((4c-1)Y + X^2 + 2c(4c-1)X + c^2(4c-1)^2)}{c(X-1+4c)(X-c+4c^2)},$$

$$x = -\frac{(X-c^2+4c^3)((4c-1)(c-1)Y - X^2 + c(4c-1)(c-3)X + c^2(c-2)(4c-1)^2)^2}{c^2(X-1+4c)^2(X-c+4c^2)^3}$$

$$y = \frac{((4c-1)(c-1)Y - X^2 + c(4c-1)(c-3)X + c^2(c-2)(4c-1)^2)^4}{c^3(X-1+4c)^4(X-c+4c^2)^4}$$

of inverse

$$Y = \frac{y^2(4y - (u-1)^4)(y - (u-1)^4)(y + u(u-1)x + (u+1)(u-1)^3)}{(u-1)^{15}(x + (u-1)^2)^2},$$

$$X = -\frac{y(4y - (u-1)^4)(y + x(u-1)^2)}{(u-1)^{10}(x + (u-1)^2)},$$

we obtain the Weierstrass equation of the new fibration #8

$$Y^2 = X^3 + 6c^2 X^2 - c^3(4c-1)(c-4)X.$$

The point $A_{22} + P_3$ corresponds to the two-torsion point $T_c = (X = 0, Y = 0)$. The section $P_c = (X = 1 - 4c, Y = (4c-1)(c-1)^2)$ corresponds to $u = \infty$ so to $\Theta_{\infty,4}$. We have also the section $X = c^3(c-4), Y = c^3(c-4)(c-1)^2$ equal to $P_c + T_c$.

So, the singular fibers are $III^*(\infty, 0), I_2(1, 4, \frac{1}{4})$. The Mordell-Weil group is of rank one generated by P_c and T_c.

7 From Fibration #8 to Fibrations with Singular Fibers of Type I_4^* and Others

From fibration #8 we can obtain many fibrations mainly with singular fibers of type I_4^* and 2-neighbor method.

Let W be the singular fiber of the fibration #8 at $c = 4$. This fiber is of type I_2 with two components $W_{4,0}$ and $W_{4,1} = \Theta_{3,1}$. We consider the following divisors:

$$D = (0) + \Theta_{1,1} + 2 \sum_{i=0}^{3} \Theta_{1,6-i} + 2\,(A_{23})$$

$$\Delta = D + R + S$$

$$D' = \Theta_{-1,1} + \Theta_{-1,3} + 2\Theta_{-1,2} + 2\,(P_3) + 2\Theta_{\infty,2} + 2\Theta_{\infty,3} + 2\,(A_{23})$$

$$\Delta' = D' + R + S,$$

where R ans S are two different elements chosen in the set $\{\Theta_{0,1}, \Theta_{-3,1}, \Theta_{\infty,3}, W_{4,0}\}$ for Δ and in the set $\{\Theta_{0,1}, \Theta_{-3,1}, \Theta_{1,3}, W_{4,0}\}$ for Δ'.

The divisors Δ and Δ' are of type I_4^*. The corresponding fibrations can be obtained with 2-neighbors method from fibration #8, and an elliptic parameter of the shape $\frac{X+k(c)}{F(c)}$, where $F(c) = \prod c^{a_1}(c-1)^{b_1}(4c-1)^{c_1}(c-4)^{d_1}$, $a_1, b_1, c_1, d_1 \in \{0, 1\}$.

To obtain $k(c)$ we use the following remarks. If R and S are different from $\Theta_{0,1}$ and $W_{4,0}$, then $(P_3 + A_{22}) \cdot \Delta$ or $(P_3 + A_{22}) \cdot \Delta' = 0$. Then the two-torsion point $T_c = (P_3 + A_{22})$ of the fibration #8 is a part of an other singular fiber of the new fibration, and we can choose $k\,(c) = 0$. The same remark is available if we consider the divisor D, R, and S different from $\Theta_{\infty,3}$. Instead of T_c we use the point $P_c(= \Theta_{\infty,4})$ and we can choose $k\,(c) = -1+4c$. This method is also true for the fibration #11 and the point $P_c + T_c$.

For the remaining cases we give a sketch of the method.

We look at the singular fibers at $c = 1, 4, \frac{1}{4} \ldots$. The singular point ω_c for this fibers is obtained at $(0, 0)$ if $c = 4$ and $\frac{1}{4}$ and at $(-3, 0)$ if $c = 1$. For the new parameter $\frac{X+k(c)}{F(c)}$, we want to obtain simple poles at the 0-components for the singular fiber at $c = c_0$. For this we do a translation of the origin $X = s + U$, $Y = w + V$ with $\omega_{c_0} \equiv (s, t) \bmod (c - c_0)$. The 0-component corresponds to $(U, V) \neq (0, 0)$, the parameter $\frac{U}{c-c_0} = \frac{X-s}{c-c_0}$ has then the good property. For the fibration #2 we can choose $s = -3$ or $-3c$.

Using Chinese remainder theorem we can obtain a translation for two different singular fibers, for example, $s = 1 - 4c = -3 - 4\,(c-1)$ gives the parameter $\frac{X-1+4c}{(c-4)(4c-1)}$ for the fibration #12; also $s = c^2 - 4c^3 = -3 - (c-1)\,(4c^2 + 3c + 3)$ for the fibration #11.

The results are given in Table 6.

7.1 From Fibration #8 to Fibration #33

We can draw two singular fibers of type IV^* and I_{10} of the same elliptic fibration #33 on the graph G_r, namely

$$D = \Theta_{-1,5} + \Theta_{-1,4} + \Theta_{-1,3} + \Theta_{-1,2} + (P_3) + \Theta_{\infty,2} + \Theta_{\infty,3} + (A_{23}) + \Theta_{0,1}$$
$$\qquad + (P_3 + A_{22})$$

$$D' = 3\Theta_{1,0} + 2\Theta_{1,1} + (A_2 + 2P_3) + 2\,(0) + \Theta_{\infty,0} + 2\Theta_{1,5} + \Theta_{1,4}.$$

Table 6 From fibration #8

Fibration	Parameter	Δ
#22 $- n$ $I_4^*(\infty, 0)$, $I_2(-1)$, $2I_1$	$\dfrac{X}{c(4c-1)}$	$D + \Theta_{-3,1} + \Theta_{\infty,3}$
#22b $- n'$ $I_4^*(\infty, 0)$, $I_2(-1)$, $2I_1$	$\dfrac{X}{c^2(4c-1)}$	$D' + \Theta_{-3,1} + \Theta_{\infty,3}$
#12 $- l$ $I_4^*(\infty)$, $I_3(0)$, $III^*(-1)$, $I_2(1/3)$	$\dfrac{X-1+4c}{(c-1)(4c-1)}$	$D + \Theta_{0,1} + \Theta_{-3,1}$
#23 $- g$ $I_4^*(\infty)$, $I_2^*(0)$, $I_2(4,-1,3)$	$\dfrac{X+3c^2}{c(c-1)}$ $= \dfrac{X+3c}{c(c-1)} + 1$	$D + \Theta_{0,1} + \Theta_{\infty,3}$
#11 $- i$ $I_4^*(\infty)$, $III^*(0)$, $I_2\left(-1, \frac{1}{3}\right)$, $I_1(-16)$	$\dfrac{X-c^2+4c^3}{c^2(4c-1)(c-1)}$ $= -1 + \dfrac{X-c^3+4c^4}{c^2(4c-1)(c-1)}$	$D' + \Theta_{0,1} + \Theta_{-3,1}$
#2 $- a$ $II^*(\infty)$, $III^*(0)$, $I_3(-1)$, $I_2(4)$	$\dfrac{X+3c}{c-1}$	
#4 $- d$ $I_4^*(\infty)$, $II^*(0)$, $I_2\left(\frac{1}{3}\right)$, $2I_1$ $II^*(\infty)$, $I_4^*(0)$, $I_2(-3)$, $2I_1$	$\dfrac{X}{(c-4)(4c-1)}$ $\dfrac{X}{c^2(c-4)(4c-1)}$ or $\dfrac{X}{c}$	$D + \Theta_{-3,1} + W_{0,4}$
#24 $- \alpha$ $I_{16}(\infty)$, $I_2(0)$, $I_1\left(-4, 4, \pm i, \pm 2\sqrt{3}\right)$	$\dfrac{Y}{Xc}$	
#33 $- \beta$ $I_{10}(\infty)$, $IV^*(0)$, $6I_1$	$\dfrac{Y}{2Xc(c-1)}$	
#20 $- \lambda$ $I_8^*(\infty)$, $I_0^*(0)$, $I_2(-3)$, $2I_1$ $I_0^*(\infty)$, $I_8^*(0)$, $I_2\left(\frac{1}{3}\right)$, $2I_1$	$\dfrac{X}{c^2}$ or $\dfrac{X}{c(c-4)(4c-1)}$	
#29 $- \delta$ $I_3^*(\infty)$, $I_4(0)$, $IV^*(-1)$, $3I_1$	$\dfrac{Y-(c-1)^2X}{X-c^3(c-4)}$ $\dfrac{Y-(c-1)^2X}{c^2(X-1+4c)}$	

To write an elliptic parameter it is more easy to use the fibration #8. We can also draw the fiber I_{10} and IV^* except one component namely $(A_2 + 2P_3)$ on Figure 5. From this figure we can take as a parameter $\beta = \dfrac{sY}{Xc(c-1)} + k$, $s, k \in \mathbb{Q}$, k chosen to have $\beta\,(A_2 + 2P_3) = 0$, so $k = 0$. To simplify we choose $s = \frac{1}{2}$, then

$$\beta = \frac{-(u-1)^3 \left(y + u(u-1)x + (u+1)(u-1)^3\right)}{2\left(y + x(u-1)^2\right)\left(x + (u-1)^2\right)}$$

and we obtain a Weierstrass equation

$$E_\beta : Y^2 = X^3 + \beta^2\left(\beta^2 - 3\right)X^2 - 2\beta^4 X + \beta^4.$$

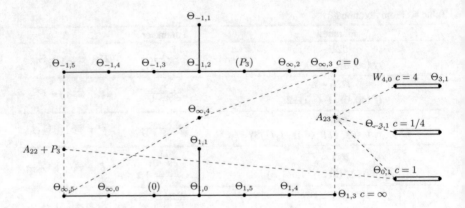

Fig. 5 Fibration #8 a new curve $W_{4,0}$

The singular fibers are

$$I_{10}(\infty), \quad IV^*(0), \quad I_1\left(12\beta^6 - 76\beta^4 + 216\beta^2 - 27\right).$$

8　From Fibration #33

For the next two paragraphs, we give a sketch of the proofs.

Let us consider the 3 sections $S = \left(0, \beta^2\right)$, $S_1 = (2\beta+1, (\beta+1)(2\beta^2-2\beta-1))$, and $S_2 = (-2\beta + 1, (-\beta + 1)(2\beta^2 + 2\beta - 1))$. The height matrix of S, S_1, S_2 has a determinant equal to $\frac{5}{2}$, so the three sections generate the Mordell-Weil group.

The following sections do not intersect the zero section, we use lines going through three of them and obtain, for example, the Weierstrass equation

$$\left(Y + \beta^2 X + 2\beta^4 - \beta^2\right)\left(Y - \beta^2 X - 2\beta^4 + \beta^2\right)$$
$$= \left(X + \beta^2\right)\left(X + 2\beta^2(\beta - 1)\right)\left(X - 2\beta^2(\beta + 1)\right)$$

$$S_1 + S_2 = \left(\tfrac{1}{4}, \tfrac{3}{4}\beta^2 - \tfrac{1}{8}\right) \qquad S+S_1 = (-2\beta^2(\beta-1), \beta^2(2\beta^3 - 4\beta^2 + 1))$$
$$S + S_1 + S_2 = \left(4\beta^2, \beta^2\left(4\beta^2 + 1\right)\right) \quad S+S_2 = (2\beta^2(\beta + 1), \beta^2(-2\beta^3 - 4\beta^2 + 1))$$
$$S_1 + S_2 + 2S = \left(-\beta^2, \beta^2(\beta^2 - 1)\right) \quad 2S = \left(3\beta^2, \beta^2(3\beta^2 - 1)\right)$$
$$3S + S_1 + S_2 = \left(4, 3\beta^2 - 8\right) \qquad 3S = \left(\tfrac{4}{9} - \tfrac{4}{3}\beta^2, \tfrac{8}{27} - \tfrac{7}{3}\beta^2 + \tfrac{4}{3}\beta^4\right).$$

Table 7 From fibration #33

Fibration	Parameter	From the form for E_β
$\#13 - t_1$ $III\,(\infty),\, I_2^*\,(0),\, I_0^*\,(1),\, I_1\left(\frac{-9}{16}\right)$	$\frac{Y-\beta^2+X\beta^2}{2X\beta^2}$	$\left(Y - \beta^2 + X\beta^2\right)\left(Y - \beta^2 - X\beta^2\right)$ $= X^2\left(X - 3\beta^2\right)$
$\#15 - \beta_0$ $III^*\,(\infty),\, I_{10}\,(0),\, 5I_1$	$\frac{Y+\beta^2X+2\beta^4-\beta^2}{\beta^2(X+2\beta^2(\beta-1))}$	$\left(Y + \beta^2 X + 2\beta^4 - \beta^2\right)$ $\left(Y - \beta^2 X - 2\beta^4 + \beta^2\right) =$ $\left(X + \beta^2\right)\left(X + 2\beta^2\,(\beta - 1)\right)$ $\left(X - 2\beta^2\,(\beta + 1)\right)$
$\#17 - q_1$ $I_{16}\,(\infty),\, IV\,(0),\, 4I_1$	$\frac{Y+\beta^2X+2\beta^4-\beta^2}{\beta(X+\beta^2)}$	$-$
$\#21 - \beta_1$ $I_6^*\,(\infty),\, I_2^*\,(0),\, I_2\left(-\frac{2}{3}\right),\, 2I_1$	$\frac{Y-\beta^2X-\beta^2}{X}$	$\left(Y - \beta^2 X - \beta^2\right)\left(Y + \beta^2 X + \beta^2\right)$ $= X\left(X + \beta^2\right)\left(X - 4\beta^2\right)$

Then we shall use a two-neighbor method and obtain the following fibrations (Table 7).

9 From Fibration #26

We start from the fibration #23 with Weierstrass equation $y^2 = x\,(x + 4t)$ $\left(x + t^2\,(t - 3)\right)$ and singular fibers of type $I_4^*\,(\infty)\, I_2^*\,(0)\,,\, 3I_2$. Using the parameter $\gamma = \frac{y}{2xt}$ we glue the two fibers $I_4^*,\, I_2^*$ to obtain a fiber I_{14}. This gives the fibration #26 and the following Weierstrass equation:

$$Y^2 + \left(4\gamma^2 + 3\right)YX - 12Y = X\,(X - 4)^2, \tag{5}$$

with singular fibers $I_{14}\,(\infty),\, I_4\,(0),\, 6I_1$.

Notice the following sections:

$P\gamma = (-4\,(2\gamma - 1)\,(\gamma + 1),\, 16\gamma\,(\gamma + 1))$ $\quad P'_\gamma = (-4\,(2\gamma + 1)\,(\gamma - 1),\, 16\gamma\,(\gamma - 1))$

$Q_\gamma = P_\gamma + P'_\gamma = (4, 0)$ $\qquad\qquad\qquad\quad 2Q_\gamma = (0, 12),\; -2Q_\gamma = (0, 0)$

$3Q_\gamma = \left(4 - 12\gamma^2,\, 16\gamma^2\,(-1 + 3\gamma^2)\right)$ $\quad 4Q_\gamma = \left(\frac{52}{9} - \frac{16}{3}\gamma^2,\, -\frac{208}{27} + \frac{64}{9}\gamma^2\right)$

$4P_\gamma + 3P'_\gamma = \left(4(2\gamma - 1)(12\gamma^3 - 24\gamma^2 + 14\gamma - 1),\right.$

$\left. 64(12\gamma^3 - 24\gamma^2 + 14\gamma - 1)(\gamma - 1)^2\gamma\right).$

The two points $P\gamma$ and $P'\gamma$ generate the Mordell-Weil group (Table 8).

The singular fiber for $\gamma = 0$ has a singular point at $(4, 0)$ and the two tangents at this point are $Y + 4(X - 4)$ and $Y - (X - 4)$.

We use the line $Y - Y_{3Q_\gamma} + 4\left(X - X_{3Q_\gamma}\right)$ or $Y - Y_{3Q_\gamma} - \left(X - X_{3Q_\gamma}\right)$. We also use the following different forms of Weierstrass equations:

Table 8 From #26

Fibration	Parameter
#5 − o	$\dfrac{3Y+4(X-4)(3\gamma^2-1)}{X-4}$
$I_{10}^*(\infty), I_2(0), I_3(-1), I_2(-16), I_1(-25)$	
#13 − t_1	$\dfrac{X}{4}$
$III(\infty), I_2^*(0), I_0^*(1), I_1\left(\frac{-9}{16}\right)$	
#14 − ρ	$\dfrac{Y+4X-16(\gamma^2-1)(3\gamma^2-1)}{\gamma(X-4+12\gamma^2)}=$
$I_{18}(\infty), 6I_1$	$\dfrac{Y-Y_{3Q_\gamma}+4\left(X-X_{3Q_\gamma}\right)}{\gamma(X-4+12\gamma^2)}$
#17 − q_1	$\dfrac{(Y+4X+16(\gamma^2-1))}{2\gamma(X-4)}$
$I_{16}(\infty), IV(0), 4I_1$	
#20 − λ	$\dfrac{Y-12}{X}$
$I_8^*(\infty), I_2(0), I_0^*(-3), 2I_1$	
#21 − β_1	$\dfrac{X-4}{\gamma^2}$
$I_2^*(\infty), I_6^*(0), I_2(-3), 2I_1$	
#32 − s	$\dfrac{3Y+13X-52+48\gamma^2}{\gamma(9X-52+48\gamma^2)}$ or $\dfrac{Y+4\gamma^2 X}{2\gamma X}$
$I_6(\infty), I_{12}(0), 6I_1$	
#38 − η	$\dfrac{(Y+4X+16(\gamma^2-1))}{2(X+4(\gamma-1)(2\gamma+1))}$
$I_2^*(\infty), I_4(0), I_8(1), 4I_1$	
#46 − β_3	$\dfrac{Y-X-4(2\gamma+1)(4\gamma^3-4\gamma^2+2\gamma-1)}{\gamma(X+4(\gamma-1)(2\gamma+1))}=$
$I_{17}(\infty), 7I_1$	$\dfrac{Y-Y_{-3Q_\gamma}-(X-X_{-3Q_\gamma})}{\gamma\left(X-X_{-3Q_\gamma}\right)}$
#47 − f_1	$\dfrac{Y-(8\gamma^2-12\gamma+1)X-4(2\gamma+1)(12\gamma^3-24\gamma^2+14\gamma-1)}{\gamma(X-4(2\gamma-1)(12\gamma^3-24\gamma^2+14\gamma-1))}$
$I_5(\infty), I_{13}(0)$	
#48 − f_2	$\dfrac{(Y+4X+16(\gamma^2-1))}{2\gamma(X+4(\gamma+1)(2\gamma-1))}$
$I_7(\infty), I_{11}(0), 6I_1$	

$$\left(Y+4X+16\left(\gamma^2-1\right)\right)\left(Y+\left(4\gamma^2-1\right)(X-4)\right)=\left(X-X_{Q_\gamma}\right)\left(X-X_{P_\gamma}\right)\left(X-X_{P_\gamma'}\right)$$
$$(Y+4\gamma X)(Y+3X-2)=\left(X-X_{Q_\gamma}\right)\left(X-X_{2Q_\gamma}\right)\left(X-X_{3Q_\gamma}\right)$$
$$(3Y+13X-52+48\gamma^2)(3Y+4(X-4)(3\gamma^2-1))=9\left(X-X_{Q_\gamma}\right)$$
$$\left(X-X_{3Q_\gamma}\right)\left(X-X_{4Q_\gamma}\right)$$
$$\left(Y-(8\gamma^2-12\gamma+1)X-4(2\gamma+1)(12\gamma^3-24\gamma^2+14\gamma-1)\right)$$
$$\left(Y+(12\gamma^2-12\gamma+4)X+16(\gamma-1)(2\gamma-1)(3\gamma^2-1)\right)$$
$$=\left(X-X_{4P_\gamma+3P_\gamma'}\right)\left(X-X_{P_\gamma}\right)\left(X-X_{3Q_\gamma}\right).$$

10 Fibration of Rank 0

We gather in a table the fibrations with rank 0 (Table 9).

Table 9 Rank 0

Fibrations	From	Parameter	Equations
#30 − e	#50 − u	$\frac{4x+(u^2-1)^2}{x(u^2-1)}$	$y^2 = x^3 + e\,(e+2)$
$I_{12}\,(\infty)\,,\,I_1^*\,(0)\,,\,I_2\,(-3,-1)\,,\,I_1\,(-4)$			$(e^2+4e+1)\,x^2 + e^2 x$
#28 − b	#50 − u	$\frac{y}{(u^2-1)^2}$	$y^2 + 6byx - 2b^2\,(b-1)^2\,y = x^3$
$IV^*\,(\infty,0)\,,\,I_6\,(1)\,,\,I_2\,(-1)$			
#12 − l	#8 − c	$\frac{x-c+4c^2}{(4c-1)(c-1)}$	$y^2 = x^3 + 2l^2\,(2l-3)\,x^2$
$I_4^*\,(\infty)\,,\,III^*\,(0)\,,\,I_3\,(1)\,,\,I_2\left(\frac{4}{3}\right)$			$-l^3\,(3l-4)\,x$
#2 − a	#8 − c	$\frac{x+3c}{c-1}$	$y^2 = x^3 - 3a^3\,(5a-4)\,x$
$II^*\,(\infty)\,,\,III^*\,(0)\,,\,I_3\,(-1)\,,\,I_2\,(4)$			$-2a^5\,(2a^2-5a+20)$
#5 − o	#2 − a	$\frac{x}{a^2}$	$y^2 = x^3 + \left(o^3 - 15o + 10\right)x^2$
$I_{10}^*\,(\infty)\,,\,I_3\,(3)\,,\,I_2\left(\frac{10}{3},-2\right)\,,\,I_1\,(-5)$			$+(-48o+160)\,x$
#10 − q	#12 − l	$\frac{x}{l^2}$	$y^2 = x^3 + q\,(q^2-6q-3)x^2 + 16q^3 x$
$I_6^*\,(\infty)\,,\,I_2^*\,(0)\,,\,I_3\,(1)\,,\,I_1\,(9)$			

11 Pairs of Fibrations with the Same Trivial Lattice

Besides the pairs #22,#22-bis and #40,#40-bis studied in [1] and [3] there are 4 other pairs of fibrations with the same trivial lattice. We need to match each Weierstrass equation with an embedding in a Niemeier lattice coming from Table 1 of [1]. In Table 10 we notice in the second column the fibration which we start. In the third column we have an elliptic parameter which allows to compute a Weierstrass equation. Then with the following remarks we can associate it to a fibration from [1]. The result is in the first column with equation given in Tables 1 and 2.

For the fibrations of parameter μ and ν, we have a two torsion section for ν and not for μ. So from [1] Table 1 the parameter μ corresponds to #19 and ν to #3.

For the fibrations of parameter π and τ, we have a two torsion section for parameter π and not for τ. So from [1] Table 1 the parameter π corresponds to #7 and τ to #6.

We look now to parameter β_1 and s_1. From the Weierstrass equations and for the singular fiber I_2 we see that the two-torsion point reduces for the case β_1 on the singular point and not in the case s_1.

For the fibration #21, from [1], we have the following primitive embedding i of $A_5 \oplus A_1$ in the Niemeier lattice L with $L_{root} = \mathbb{D}_{12}^2 = \mathbb{D}_{12}^{(1)} \times \mathbb{D}_{12}^{(2)}$:

$$i(A_5) = \langle d_{12}^{(1)}, d_{10}^{(1)}, d_9^{(1)}, d_8^{(1)}, d_7^{(1)} \rangle$$

$$i(A_1) = \langle d_{12}^{(2)} \rangle.$$

Let N be the orthogonal complement of $i\,(A_5 \oplus A_1)$ in L_{root}, we have then $N_{root} = \mathbb{D}_6 \mathbb{D}_{10} \mathbb{A}_1$, where $\mathbb{A}_1 = \langle d_{11}^{(2)} \rangle$. We can construct the element of 2-torsion in the Mordell-Weil group

Table 10 Pairs of fibrations

Fibration	From	Parameter
#3 − ν $I_{12}^*(\infty)$, $I_2(0)$, $I_1(-1,15,\nu_1,\overline{\nu_1})$	#1 − h $y^2=x^3-75xh+2h^5\left(h^2-121h+4\right)$	$\nu=\frac{x}{h^2}+5$
#19 − μ $I_{12}^*(\infty)$, $I_2\left(\frac{4}{3}\right)$, $4I_1$	#4 − d	$\mu=\frac{x+4d^3+3d^2}{d^2}$
#7 − π $III^*(\infty)$, $I_6^*(0)$, $I_1\left(\frac{1}{4},\pi_1,\overline{\pi_1}\right)$	#3 − ν	$\pi=\frac{x+4}{\nu}$
#6 − τ $III^*(\infty)$, $I_6^*(0)$, $3I_1$	#19 − μ	$\tau=\frac{x-3}{3\mu-4}$
#9 − s_1 $I_6^*(\infty)$, $I_2^*(0)$, $I_2(-4)$, $I_1(-1,-9)$	#11 − i	$s_1=\frac{x}{(i-1)^2}$
#21 − β_1 $I_6^*(\infty)$, $I_2^*(0)$, $I_2\left(-\frac{2}{3}\right)$, $2I_1$	#33 − β	$\beta_1=\frac{Y-\beta^2X-\beta^2}{X}$
#27 − f_4 $I_{10}(\infty)$, $I_3^*(0)$, $I_2(3)$, $3I_1$	#48 − f_2	$f_4=\frac{y+xf_2(f_2-1)}{f_2x}$
#31 − σ_2 $I_3^*(\infty)$, $I_{10}(0)$, $I_2(2)$, $3I_1$	#16 − σ_1	$\sigma_2=\frac{y-4}{x-2\sigma_1-6}$

$$O+2F+\left(\delta_{12},\bar{\delta}_{12}\right).$$

Indeed we have $2\delta_{12}^{(1)}=z_6+u$ with $u\in\mathbb{D}_6$, and $2\bar{\delta}_{12}$ in $\mathbb{D}_{10}$. Moreover we have $d_{11}^{(2)}\cdot\left(\delta_{12},\bar{\delta}_{12}\right)=1$. The last equality means that the two torsion section and the zero section intersect two different component of the singular fiber I_2. This is the case for the fibration with the parameter β_1.

For the last parameters f_4 and σ_2 we consider the singular fiber I_2. From the Weierstrass equations we consider a generator of the Mordell-Weil group. For the parameter f_4 the section reduces on the singular point; this is not the case for the parameter σ_2.

Recall the primitive embedding i of $A_5\oplus A_1$ in the Niemeier lattice L with $L_{root}=\mathbb{D}_9\mathbb{A}_{15}$:

$$i\,(\mathbb{A}_5)=\langle a_1,a_2,a_3,a_4,a_5\rangle$$
$$i\,(\mathbb{A}_1)=\langle d_9\rangle.$$

In this case, $N_{root}=\mathbb{A}_1\mathbb{D}_7\mathbb{A}_9$. The Mordell-Weil group is generated by the section

$$O+kF+\left(\delta_9,2\alpha_{15}-a_1+a_3+2a_4+3a_5+4a_6\right),$$

where $\alpha_{15}=\varepsilon_1-\frac{1}{16}\sum_{j=1}^{16}\varepsilon_j=\frac{1}{16}\sum_{j=1}^{15}(15-j+1)\,a_j$.

Table 11 Last fibrations

Fibration	From	Parameter
$\#16 - \sigma_1$		
$I_{16}(\infty), I_2(0), 6I_1$	$\#9 - s_1$	$\sigma_1 = \frac{y}{xs_1}$
$\#18 - \delta_2$		
$I_{12}(\infty), I_2^*(0), 4I_1$	$\#27 - f_4$	$\delta_2 = \frac{y}{xf_4}$
$\#25 - m_1$		
$I_5^*(\infty), I_8(0), 5I_1$	$\#34 - m$	$m_1 = \frac{x}{m^2}$

The orthogonal complement of $i\,(\mathbb{A}_1)$ in $\mathbb{D}_9$ is equal to $\langle d_8 \rangle \oplus \mathbb{D}_7$, and we have $\delta_9 \cdot d_8 \neq 0$. So the fibration corresponds to the fibration of parameter f_4.

12 The Last One

See Table 11.

References

1. M.J. Bertin, A. Garbagnati, R. Hortsch, O. Lecacheux, M. Mase, C. Salgado, U. Whitcher, *Classifications of Elliptic Fibrations of a Singular K3 Surface*, in Women in Numbers Europe, 17–49, Research Directions in Number Theory, Association for Women in Mathematics Series, Springer, 2015.
2. M.J. Bertin, O. Lecacheux, *Elliptic fibrations on the modular surface associated to $\Gamma_1(8)$*, in Arithmetic and geometry of K3 surfaces and Calabi-Yau threefolds, 153–199, Fields Inst. Commun., **67**, Springer, New York, 2013.
3. M.J. Bertin, O. Lecacheux, *Automorphisms of certain Niemeier lattices and Elliptic Fibrations,*Albanian Journal of Mathematics Volume 11, Number 1,(2017) 13–34.
4. N. Elkies, A. Kumar, *K3 surfaces and equations for Hilbert modular surfaces,* Algebra Number Theory, Volume 8, Number 10 (2014), 2297–2411.
5. T. Harrache, O. Lecacheux, *Étude des fibrations elliptiques d'une surface K3,* Journal de théorie des nombres de Bordeaux, Tome 23 (2011) 1, 183–207.
6. A. Garbagnati, A. Sarti, *Elliptic fibrations and symplectic automorphisms on K3 surfaces,* Comm. in Algebra 37 (2009), 10,3601–3631.
7. A. Kumar, *Elliptic fibrations on a generic Jacobian Kummer surface* J. Algebraic Geometry 23 (2014), 599–667.
8. M. Kuwata, Maple Library Elliptic Surface Calculator, http://c-faculty.chuo-u.ac.jp/~kuwata/ESC.php.
9. K. Nishiyama, *The Jacobian fibrations on some K3 surfaces and their Mordell-Weil groups,* Japan. J. Math. (N.S.) **22** (1996) 293–347.
10. I.-I. Piatetski-Shapiro , I.-R., Shafarevich, *Torelli's theorem for algebraic surfaces of type K3,* Izv. Akad. Nauk SSSR Ser. Mat. **35**, (1971) 530–572.
11. T. Sengupta *Elliptic fibrations on supersingular K3 surface with Artin invariant 1 in characteristic 3.* preprint, 2012, arXiv:1204.6478v2.
12. M. Schütt, T. Shioda, *Elliptic surfaces,* Algebraic geometry in East Asia–Seoul 2008, Adv. Stud. Pure Math., **60**, Math. Soc. Japan, Tokyo, (2010) 51–160.

13. T. Shioda, *On elliptic modular surfaces*, J. Math. Soc. Japan 24 (1972), 20–59.
14. T. Shioda, *On the Mordell-Weil lattices*, Comment. Math. Univ. St. Paul. 39 (1990), no. 2, 211–240.
15. J. H. Silverman, *Computing heights on elliptic curves*, Math. Comp. **51** (183) (1988) 339–358.
16. J. Silverman, J. Tate *Rational Points on Elliptic Curves* Undergraduate Texts in Mathematics book series (UTM) 1992.
17. K. Utsumi *Weierstrass equations for Jacobian fibrations on a certain K3 surface* Hiroshima Math. J. **42** (2012), 355–383.

Newton Polygons of Cyclic Covers of the Projective Line Branched at Three Points

Wanlin Li, Elena Mantovan, Rachel Pries, and Yunqing Tang

Abstract We review the Shimura–Taniyama method for computing the Newton polygon of an abelian variety with complex multiplication. We apply this method to cyclic covers of the projective line branched at three points. As an application, we produce multiple new examples of Newton polygons that occur for Jacobians of smooth curves in characteristic p. Under certain congruence conditions on p, these include: the supersingular Newton polygon for each genus g with $4 \leq g \leq 11$; nine non-supersingular Newton polygons with p-rank 0 with $4 \leq g \leq 11$; and, for all $g \geq 5$, the Newton polygon with p-rank $g - 5$ having slopes $1/5$ and $4/5$.

Keywords Curve · Cyclic cover · Jacobian · Abelian variety · Moduli space · Reduction · Supersingular · Newton polygon · p-rank · Dieudonné module · p-divisible group · Complex multiplication · Shimura–Taniyama method

MSC10 Primary 11G20, 11M38, 14G10, 14H40, 14K22; Secondary 11G10, 14H10, 14H30, 14H40

W. Li (✉)
Department of Mathematics, University of Wisconsin, Madison, WI 53706, USA
e-mail: wanlin@math.wisc.edu

E. Mantovan
Department of Mathematics, California Institute of Technology, Pasadena, CA 91125, USA
e-mail: mantovan@caltech.edu

R. Pries
Department of Mathematics, Colorado State University, Fort Collins, CO 80523, USA
e-mail: pries@math.colostate.edu

Y. Tang
Department of Mathematics, Princeton University, Princeton, NJ 08540, USA
e-mail: yunqingt@math.princeton.edu

© The Author(s) and The Association for Women in Mathematics 2019
J. S. Balakrishnan et al. (eds.), *Research Directions in Number Theory*, Association for Women in Mathematics Series 19, https://doi.org/10.1007/978-3-030-19478-9_5

1 Introduction

In positive characteristic p, there are several discrete invariants associated with abelian varieties, e.g., the p-rank, the Newton polygon, and the Ekedahl–Oort type. These invariants give information about the Frobenius morphism and the number of points of the abelian variety defined over finite fields. It is a natural question to ask which of these invariants can be realized by Jacobians of smooth curves.

For any prime p, genus g, and f such that $0 \leq f \leq g$, Faber and van der Geer prove in [4] that there exists a smooth curve of genus g defined over $\overline{\mathbb{F}}_p$ which has p-rank f. Much less is known about the Newton polygon, more precisely, the Newton polygon of the characteristic polynomial of Frobenius. For $g = 1, 2, 3$, it is known that every possible Newton polygon occurs for a smooth curve of genus g. Beyond genus 3, very few examples of Newton polygons are known to occur. In [11, Expectation 8.5.4], for $g \geq 9$, Oort observed that it is unlikely for all Newton polygons to occur for Jacobians of smooth curves of genus g.

This project focuses on Newton polygons of cyclic covers of the projective line $\mathbb{P}^1$. One case which is well-understood is when the cover is branched at 3 points, especially when the cover is of prime degree (see [6, 7, 16, 18]). In this case, the Jacobian is an abelian variety with complex multiplication and its Newton polygon can be computed using the Shimura–Taniyama theorem. In Section 3, we give a survey of this material, including composite degree. Although this material is well-known, it has not been systematically analyzed for this application. We use this method to tabulate numerous Newton polygons having p-rank 0 which occur for Jacobians of smooth curves.

We now describe the main results of this paper in more detail. By [15, Theorem 2.1], if $p = 2$ and $g \in \mathbb{N}$, then there exists a supersingular curve of genus g defined over $\overline{\mathbb{F}}_2$. For this reason, we restrict to the case that p is odd in the following result. In the first application, we verify the existence of supersingular curves in the following cases. In Remark 5.2, we explain why the $g = 9$, $g = 10$, and $g = 11$ cases are especially interesting.

Theorem 1.1 (See Theorem 5.1). *Let p be odd. There exists a smooth supersingular curve of genus g defined over $\overline{\mathbb{F}}_p$ in the following cases:*

$g = 4$ *and* $p \equiv 2 \bmod 3$, *or* $p \equiv 2, 3, 4 \bmod 5$;

$g = 5$ *and* $p \equiv 2, 6, 7, 8, 10 \bmod 11$;

$g = 6$ *and* $p \not\equiv 1, 3, 9 \bmod 13$ *or* $p \equiv 3, 5, 6 \bmod 7$;

$g = 7$ *and* $p \equiv 14 \bmod 15$ *or* $p \equiv 15 \bmod 16$;

$g = 8$ *and* $p \not\equiv 1 \bmod 17$;

$g = 9$ *and* $p \equiv 2, 3, 8, 10, 12, 13, 14, 15, 18 \bmod 19$;

$g = 10$ *and* $p \equiv 5, 17, 20 \bmod 21$;

$g = 11$ *and* $p \equiv 5, 7, 10, 11, 14, 15, 17, 19, 20, 21, 22 \bmod 23$.

The second application is Theorem 5.4: under certain congruence conditions on p, we prove that nine new Newton polygons that have p-rank 0 but are not supersingular occur for Jacobians of smooth curves.

For context for the third application, recall that every abelian variety is isogenous to a factor of a Jacobian. This implies that every rational number $\lambda \in [0, 1]$ occurs as a slope for the Newton polygon of the Jacobian of a smooth curve. In almost all cases, however, there is no control over the other slopes in the Newton polygon.

In the third application, when $d = 5$ or $d = 11$, under congruence conditions on p, we show that the slopes $1/d$ and $(d - 1)/d$ occur for a smooth curve in characteristic p of arbitrarily large genus with complete control over the other slopes in the Newton polygon. Namely, under these congruence conditions on p and for all $g \geq d$, we prove that there exists a smooth curve of genus g defined over $\overline{\mathbb{F}}_p$ whose Newton polygon contains the slopes $1/d$ and $(d - 1)/d$ with multiplicity d and slopes 0 and 1 with multiplicity $g - d$. This was proven earlier, for all p, when $d = 2$ [4, Theorem 2.6]; $d = 3$ [13, Theorem 4.3]; and $d = 4$ [2, Corollary 5.6].

Let $G_{1,d-1} \oplus G_{d-1,1}$ denote the p-divisible group with slopes $1/d, (d - 1)/d$.

Theorem 1.2 (See Theorem 5.5). *For the following values of d and p and for all $g \geq d$, there exists a smooth curve of genus g defined over $\overline{\mathbb{F}}_p$ whose Jacobian has p-divisible group isogenous to $(G_{1,d-1} \oplus G_{d-1,1}) \oplus (G_{0,1} \oplus G_{1,0})^{g-d}$:*

(1) $d = 5$ *for all* $p \equiv 3, 4, 5, 9 \bmod 11;$
(2) $d = 11$ *for all* $p \equiv 2, 3, 4, 6, 8, 9, 12, 13, 16, 18 \bmod 23$.

In future work, we determine new results about Newton polygons of curves arising in positive-dimensional special families of cyclic covers of the projective line. This work relies on the Newton polygon stratification of PEL-type Shimura varieties. Then we attack the same questions for arbitrarily large genera using a new induction argument for Newton polygons of cyclic covers of $\mathbb{P}^1$. We use the Newton polygons found in this paper as base cases in this induction process.

Organization of the Paper

Section 2 contains basic definitions and facts about group algebras, cyclic covers of $\mathbb{P}^1$, and Newton polygons.

Section 3 focuses on the Jacobians of cyclic covers branched at exactly three points. We review the Shimura–Taniyama method for computing the Newton polygon and provide examples.

Section 4 contains tables of data.

Section 5 contains the proofs of the three theorems.

Acknowledgements.

This project began at the *Women in Numbers 4* workshop at the Banff International Research Station. Pries was partially supported by NSF grant DMS-15-02227. We thank the referee for the valuable feedback and comments.

2 Notation and Background

2.1 The Group Algebra $\mathbb{Q}[\mu_m]$

For an integer $m \geq 2$, let $\mu_m := \mu_m(\mathbb{C})$ denote the group of m-th roots of unity in $\mathbb{C}$. For each positive integer d, we fix a primitive d-th root of unity $\zeta_d = e^{2\pi i/d} \in \mathbb{C}$. Let $K_d = \mathbb{Q}(\zeta_d)$ be the d-th cyclotomic field over $\mathbb{Q}$ of degree $\phi(d)$.

Let $\mathbb{Q}[\mu_m]$ denote the group algebra of μ_m over $\mathbb{Q}$. It has an involution $*$ induced by the inverse map on μ_m, i.e., $\zeta^* := \zeta^{-1}$ for all $\zeta \in \mu_m$. The $\mathbb{Q}$-algebra $\mathbb{Q}[\mu_m]$ decomposes as a product of finitely many cyclotomic fields, namely

$$\mathbb{Q}[\mu_m] = \prod_{0 < d \mid m} K_d.$$

The involution $*$ on $\mathbb{Q}[\mu_m]$ preserves each cyclotomic factor K_d, and for each $d \mid m$, the restriction of $*$ to K_d agrees with complex conjugation.

Let $\mathcal{T}$ denote the set of homomorphisms $\tau : \mathbb{Q}[\mu_m] \to \mathbb{C}$. In the following, we write

$$\mathbb{Q}[\mu_m] \otimes_{\mathbb{Q}} \mathbb{C} = \prod_{\tau \in \mathcal{T}} \mathbb{C},$$

and for each $(\mathbb{Q}[\mu_m] \otimes_{\mathbb{Q}} \mathbb{C})$-module W, we write $W = \oplus_{\tau \in \mathcal{T}} W_\tau$, where W_τ denotes the subspace of W on which $a \otimes 1 \in \mathbb{Q}[\mu_m] \otimes_{\mathbb{Q}} \mathbb{C}$ acts as $\tau(a)$.

For convenience, we fix an identification $\mathcal{T} = \mathbb{Z}/m\mathbb{Z}$ by defining, for $n \in \mathbb{Z}/m\mathbb{Z}$,

$$\tau_n(\zeta) := \zeta^n, \quad \text{for all } \zeta \in \mu_m.$$

Note that, for any $n \in \mathbb{Z}/m\mathbb{Z}$, and $a \in \mathbb{Q}[\mu_m]$,

$$\tau_{-n}(a) = \tau_n(a^*) = \overline{\tau_n(a)},$$

where $z \mapsto \bar{z}$ denotes complex conjugation on $\mathbb{C}$. In the following, we write $\tau_n^* := \tau_{-n}$.

Remark 2.1. For each $\tau \in \mathcal{T}$, the homomorphism $\tau : \mathbb{Q}[\mu_m] \to \mathbb{C}$ factors via a projection $\mathbb{Q}[\mu_m] \to K_d$, for a unique positive divisor d of m. We refer to d as *the order of* τ. Indeed, for each $n \in \mathbb{Z}/m\mathbb{Z}$, the homomorphism τ_n factors via the cyclotomic field K_d if and only if d is the exact order of n in $\mathbb{Z}/m\mathbb{Z}$.

For each rational prime p, we fix an algebraic closure $\mathbb{Q}_p^{\mathrm{alg}}$ of $\mathbb{Q}_p$, and an identification $\mathbb{C} \simeq \mathbb{C}_p$, where $\mathbb{C}_p$ denotes the p-adic completion of $\mathbb{Q}_p^{\mathrm{alg}}$. We denote by $\mathbb{Q}_p^{\mathrm{un}}$ the maximal unramified extension of $\mathbb{Q}_p$ in $\mathbb{Q}_p^{\mathrm{alg}}$, and by σ the Frobenius of $\mathbb{Q}_p^{\mathrm{un}}$.

Assume that p does not divide m. Then $\mathbb{Q}[\mu_m]$ is unramified at p (i.e., the group μ_m is étale over $\mathbb{Z}_p$), and, for each $\tau \in \mathcal{T}$, the homomorphism $\tau : \mathbb{Q}[\mu_m] \to \mathbb{C} \simeq \mathbb{C}_p$ factors via the subfield $\mathbb{Q}_p^{\text{un}} \subset \mathbb{C}_p$. In particular,

$$\mathbb{Q}[\mu_m] \otimes_{\mathbb{Q}} \mathbb{Q}_p^{\text{un}} = \prod_{\tau \in \mathcal{T}} \mathbb{Q}_p^{\text{un}}.$$

There is a natural action of the Frobenius σ on the set $\mathcal{T}$, defined by $\tau \mapsto \tau^\sigma :=$ $\sigma \circ \tau$. Then $\tau_n^\sigma = \tau_{pn}$ for all $n \in \mathbb{Z}/m\mathbb{Z}$.

We write $\mathfrak{O}$ for the set of σ-orbits $\mathfrak{o}$ in $\mathcal{T}$. For each $\tau \in \mathcal{T}$, we denote by $\mathfrak{o}_\tau$ its σ-orbit. The set $\mathfrak{O}$ is in one-to-one correspondence with the set of primes $\mathfrak{p}$ of $\mathbb{Q}[\mu_m]$ above p. We write $\mathfrak{p}_\mathfrak{o}$ for the prime above p associated with an orbit $\mathfrak{o}$ in $\mathcal{T}$. For each σ-orbit $\mathfrak{o} \in \mathcal{T}$, the order of τ is the same for all $\tau \in \mathfrak{o}$ and we denote this order by $d_\mathfrak{o}$. Let $K_{d_\mathfrak{o},\mathfrak{p}_\mathfrak{o}}$ denote the completion of $K_{d_\mathfrak{o}}$ along the prime $\mathfrak{p}_\mathfrak{o}$. Then

$$\mathbb{Q}[\mu_m] \otimes_{\mathbb{Q}} \mathbb{Q}_p = \prod_{\mathfrak{o} \in \mathfrak{O}} K_{d_\mathfrak{o},\mathfrak{p}_\mathfrak{o}}.$$

2.2 Cyclic Covers of the Projective Line

Fix an integer $m \geq 2$, together with a triple of positive integers $a = (a(1), a(2), a(3))$. We refer to such a pair (m, a) as a *monodromy datum* if

(1) $a(i) \not\equiv 0 \bmod m$, for all $i = 1, 2, 3$,
(2) $\gcd(m, a(1), a(2), a(3)) = 1$,
(3) $a(1) + a(2) + a(3) \equiv 0 \bmod m$.

Fix a monodromy datum (m, a). The equation

$$y^m = x^{a(1)}(x - 1)^{a(2)} \tag{2.1}$$

defines a smooth projective curve $C = C_{(m,a)}$ defined over $\mathbb{Q}$. The function x on C yields a map $C \to \mathbb{P}^1$, and there is a μ_m-action on C over $\mathbb{P}^1$ given by $\zeta \cdot (x, y) = (x, \zeta \cdot y)$ for all $\zeta \in \mu_m$ (more precisely, this action is defined on the base change of C from $\mathbb{Q}$ to K_m). The curve C, together with this μ_m-action, is a μ_m-Galois cover of the projective line $\mathbb{P}^1$; it is branched at $0, 1, \infty$ and has local monodromy $a(1)$ at 0, $a(2)$ at 1, and $a(3)$ at ∞. By the hypotheses on the monodromy datum, for primes $p \nmid m$ the reduction of C at p is a geometrically irreducible curve of genus g, where

$$g = g(m, a) = 1 + \frac{m - \gcd(a(1), m) - \gcd(a(2), m) - \gcd(a(3), m)}{2}. \tag{2.2}$$

Remark 2.2. The isomorphism class of the curve $C = C_{(m,a)}$ depends only on the equivalence class of the monodromy datum (m, a), where two monodromy data

(m, a) and (m', a') are equivalent if $m = m'$, and the images of a, a' in $(\mathbb{Z}/m\mathbb{Z})^3$ are in the same orbit under the action of $(\mathbb{Z}/m\mathbb{Z})^* \times \Sigma_3$, where Σ_3 is the symmetric group of degree 3.

Let $V := H^1(C(\mathbb{C}), \mathbb{Q})$ denote the first Betti cohomology group of C. Then V is a $\mathbb{Q}[\mu_m]$-module, and there is a decomposition $V \otimes_{\mathbb{Q}} \mathbb{C} = \oplus_{\tau \in \mathcal{T}} V_\tau$. In addition, V has a Hodge structure of type $(1, 0) + (0, 1)$,with the $(1, 0)$ piece given by $H^0(C(\mathbb{C}), \Omega_C^1)$ via the Betti–de Rham comparison. We denote by V^+ (resp. V^-) the $(1, 0)$ (resp. $(0, 1)$) piece. Both V^+ and V^- are $\mathbb{Q}[\mu_m]$-modules, so there are decompositions

$$V^+ = \oplus_{\tau \in \mathcal{T}} V_\tau^+ \text{ and } \quad V^- = \oplus_{\tau \in \mathcal{T}} V_\tau^-.$$

Let $\mathfrak{f}(\tau) := \dim_{\mathbb{C}} V_\tau^+$. For any $q \in \mathbb{Q}$, let $\langle q \rangle$ denote the fractional part of q. By [10, Lemma 2.7, Section 3.2] (see also [3]),

$$\mathfrak{f}(\tau_n) = \begin{cases} -1 + \sum_{i=1}^3 \langle \frac{-na(i)}{m} \rangle & \text{if } n \not\equiv 0 \bmod m \\ 0 & \text{if } n \equiv 0 \bmod m. \end{cases} \tag{2.3}$$

We call $\mathfrak{f} = (\mathfrak{f}(\tau_1), \ldots, \mathfrak{f}(\tau_{m-1}))$ the *signature type* of the monodromy datum (m, a).

Remark 2.3. Let $n(\tau) := \dim_{\mathbb{C}} V_\tau$. For all $\tau \in \mathcal{T}$, one sees that $\dim_{\mathbb{C}} V_{\tau^*}^+ = \dim_{\mathbb{C}} V_\tau^-$ and thus $\mathfrak{f}(\tau) + \mathfrak{f}(\tau^*) = n(\tau)$. Note that $n(\tau)$ depends only on the order of τ, and thus only on the orbit $\mathfrak{o}_\tau$. If $\mathfrak{o} \in \mathfrak{O}$, we sometimes write $n(\mathfrak{o}) = n(\tau)$, for any $\tau \in \mathfrak{o}$.

2.3 Newton Polygons

Let X denote a g-dimensional abelian scheme over an algebraically closed field $\mathbb{F}$ of positive characteristic p.

If $\mathbb{F}$ is an algebraic closure of $\mathbb{F}_p$, the finite field of p elements, then there exists a finite subfield $\mathbb{F}_0 \subset \mathbb{F}$ such that X is isomorphic to the base change to $\mathbb{F}$ of an abelian scheme X_0 over $\mathbb{F}_0$. Let $W(\mathbb{F}_0)$ denote the Witt vector ring of $\mathbb{F}_0$. Consider the action of Frobenius φ on the crystalline cohomology group $H^1_{\text{cris}}(X_0/W(\mathbb{F}_0))$. There exists an integer n such that φ^n, the composite of n Frobenius actions, is a linear map on $H^1_{\text{cris}}(X_0/W(\mathbb{F}_0))$. The Newton polygon $\nu(X)$ of X is defined as the multi-set of rational numbers λ such that $n\lambda$ are the valuations at p of the eigenvalues of Frobenius for this action. Note that the Newton polygon is independent of the choice of X_0, $\mathbb{F}_0$, and n.

Here is an alternative definition, which works for arbitrary algebraically closed field $\mathbb{F}$. For each $n \in \mathbb{N}$, consider the multiplication-by-p^n morphism $[p^n] : X \to X$ and its kernel $X[p^n]$. The p-divisible group of X is denoted by $X[p^\infty] =$

$\varinjlim X[p^n]$. For each pair (c, d) of non-negative relatively prime integers, fix a p-divisible group $G_{c,d}$ of codimension c, dimension d, and thus height $c + d$. By the Dieudonné–Manin classification [8], there is an isogeny of p-divisible groups

$$X[p^\infty] \sim \oplus_{\lambda = \frac{d}{c+d}} G_{c,d}^{m_\lambda},$$

where (c, d) ranges over pairs of non-negative relatively prime integers. The Newton polygon is the multi-set of values of the slopes λ. By identifying $H^1_{\mathrm{cris}}(X/W(\mathbb{F}))$ with the (contravariant) Dieudonné module of X, it is possible to show that these definitions are equivalent.

The slopes of the Newton polygon are in $\mathbb{Q} \cap [0, 1]$. The Newton polygon is typically drawn as a lower convex polygon, with endpoints $(0, 0)$ and $(2g, g)$ and slopes equal to the values of λ, with multiplicity $(c + d)m_\lambda$. It is symmetric and has integral breakpoints. The Newton polygon is an isogeny invariant of A; it is determined by the multiplicities m_λ.

Given an abelian variety or p-divisible group $\mathcal{A}$ defined over a local field of mixed characteristic $(0, p)$, by abuse of notation, we may write $\nu(\mathcal{A})$ for the Newton polygon of its special fiber.

In this paper, we use *ord* to denote the Newton polygon with slopes $0, 1$ with multiplicity 1 and ss to denote the Newton polygon with slope $1/2$ with multiplicity 2. For $s < t \in \mathbb{Z}_{>0}$ with $\gcd(s, t) = 1$, we use $(s/t, (t-s)/t)$ to denote the Newton polygon with slopes s/t and $(t - s)/t$ with multiplicity t.

Definition 2.4. The *p-rank* of X is defined to be $\dim_{\mathbb{F}_p} \mathrm{Hom}(\mu_p, X)$. Equivalently, the p-rank of X is the multiplicity of the slope 0 in the Newton polygon.

Definition 2.5. Given a finite set of lower convex polygons $\{\nu_i \mid i = 1, \ldots, n\}$, $n \geq 2$, each ν_i having end points $(0, 0)$ and (h_i, d_i) and with slope λ occurring with multiplicity $m_{i,\lambda}$, their *amalgamate sum* $\sum_{i=1}^n \nu_i$ is the lower convex polygon having end points $(0, 0)$ and $(\sum_{i=1}^n h_i, \sum_{i=1}^n d_i)$ and with slope λ occurring with multiplicity $\sum_{i=1}^n m_{i,\lambda}$.

For any finite set of p-divisible groups $\{G_i \mid i = 1, \ldots, n\}$, $n \geq 2$, with Newton polygons $\nu(G_i)$, the Newton polygon of the p-divisible group $\oplus_{i=1}^n G_i$ is the amalgamate sum $\sum_{i=1}^n \nu(G_i)$.

3   Newton Polygons of Curves with Complex Multiplication

As in Section 2.2, we fix a monodromy datum (m, a), and consider the μ_m-Galois cover $C_{(m,a)} \to \mathbb{P}^1$ branched at $0, 1, \infty$ with local monodromy $a = (a(1), a(2), a(3))$ as in (2.1). We write $C = C_{(m,a)}$ and let $J = J_{(m,a)}$ be the Jacobian $\mathrm{Jac}(C)$ of C. The action of μ_m on C induces an action of $\mathbb{Q}[\mu_m]$ on J. Also, the equation of C naturally defines integral models $\mathcal{C}$ and $\mathcal{J} = \mathrm{Jac}(\mathcal{C})$ of C

and J over $\mathbb{Z}$ [17, Section 4]; the curve C has good reduction at all primes p such that $p \nmid m$ by [1, XIII, Corollary 2.12, Proposition 5.2].

3.1 Shimura–Taniyama Method

It is well-known that J is an abelian variety with complex multiplication, but we record a proof here with a refined statement on the CM algebra contained in $\operatorname{End}(J_{\mathbb{Q}^{\mathrm{alg}}}) \otimes \mathbb{Q}$.

We say that an abelian variety A over $\mathbb{Q}^{\mathrm{alg}}$ has complex multiplication (CM) by a $\mathbb{Q}$-algebra E if E is an étale $\mathbb{Q}$-subalgebra of $\operatorname{End}(A_{\mathbb{Q}^{\mathrm{alg}}}) \otimes \mathbb{Q}$ of degree $2 \dim A$ over $\mathbb{Q}$. In particular, if $E = \prod E_i$, then A has CM by E if and only if A is isogenous to $\prod A_i$ with A_i an abelian variety with CM by E_i. Also, if A has CM by E, then $H_1(A, \mathbb{Q})$ is free of rank 1 over E ([9, Definition 3.2, and Proposition 3.6]).

Lemma 3.1. *The abelian variety J has complex multiplication by $\prod_d K_d$, where the product is taken over all d such that $1 < d \mid m$ and $d \nmid a(i)$ for any $i = 1, 2, 3$.*

Proof. By Hodge theory, $J_{\overline{\mathbb{Q}}}$ is isogenous to $\prod_{1 < d \mid m} A_d$, where A_d is an abelian variety whose first Betti cohomology group is isomorphic to $\oplus_{\tau \text{ of order } d} V_\tau$.

For $0 \neq n \in \mathbb{Z}/m\mathbb{Z}$, the order d of n is $d = m/\gcd(n, m)$. Let x_n be the number of elements in a which are not divisible by d. If none of the $a(i)$ is divisible by d, then $x_n = 3$. Otherwise, $x_n = 2$ since $\gcd(m, a(1), a(2), a(3)) = 1$ and $a(1) + a(2) + a(3) \equiv 0 \bmod m$. For example, when $\gcd(n, m) = 1$, then $d = m$ and hence $x_n = 3$. For example, if m is even and $n = m/2$, then $d = 2$ and $x_n = 2$. By (2.3), $\mathfrak{f}(\tau_n) + \mathfrak{f}(\tau_{-n}) = x_n - 2$. Hence $A_d = \{1\}$, the trivial abelian variety if $x_n = 2$, and A_d is a simple abelian variety of dimension $[K_d : \mathbb{Q}]/2$ such that $\operatorname{End}(A_d) \otimes \mathbb{Q} = K_d$ otherwise.

Then J has complex multiplication by the product of the fields K_d where the product is taken over all d such that if $x_n = 3$. $\square$

By Lemma 3.1, the abelian variety J has potentially good reduction everywhere. This means that there exists an abelian scheme over some finite extension of $\mathbb{Z}_p$ such that its generic fiber is J. For $p \nmid m$, since C already has good reduction at p, so does J; no extension of $\mathbb{Z}_p$ is needed, the Jacobian $\mathcal{J}$ is a smooth integral model of J defined over $\mathbb{Z}_p$. The $\mathbb{Q}[\mu_m]$-action on J extends naturally to a $\mathbb{Q}[\mu_m]$-action on $\mathcal{J}$.

Let $\mathcal{J}[p^\infty]$ be the associated p-divisible group scheme of $\mathcal{J}$. The $\mathbb{Q}[\mu_m]$-action on $\mathcal{J}$ induces a $(\mathbb{Q}[\mu_m] \otimes_{\mathbb{Q}} \mathbb{Q}_p)$-action on $\mathcal{J}[p^\infty]$ and thus a canonical decomposition

$$\mathcal{J}[p^\infty] = \oplus_{\mathfrak{o} \in \mathfrak{O}'} \mathcal{J}[\mathfrak{p}_\mathfrak{o}^\infty],$$

where $\mathfrak{O}'$ is the subset of $\mathfrak{O}$ with $d_\mathfrak{o} \nmid a(i)$ for any $i = 1, 2, 3$ and each p-divisible group $\mathcal{J}[\mathfrak{p}_\mathfrak{o}^\infty]$ has height #$\mathfrak{o}$.

To state the theorem by Shimura–Taniyama and Tate on $\nu(\mathcal{J}[\mathfrak{p}_0^\infty])$, we introduce the following notation. Recall that $\mathfrak{f}(\tau) = \dim_{\mathbb{C}} V_\tau^+$. By the proof of Lemma 3.1, $\mathfrak{f}(\tau) \in \{0, 1\}$ and $\mathfrak{f}(\tau) + \mathfrak{f}(\tau^*) = 1$ for all $\tau \in \mathcal{T}$ such that the order of τ does not divide $a(i)$ for any $i = 1, 2, 3$. For $\epsilon \in \{0, 1\}$, define

$$S_\epsilon = \{\tau \in \mathcal{T} \mid \text{the order of } \tau \text{ does not divide } a(1), a(2), a(3), \text{ and } \mathfrak{f}(\tau) = \epsilon\}.$$
(3.1)

For $\mathfrak{o} \in \mathfrak{O}'$, set $\alpha_\mathfrak{o} = \#(\mathfrak{o} \cap S_1)$ and $\beta_\mathfrak{o} = \#(\mathfrak{o} \cap S_0)$. Note that $\alpha_\mathfrak{o} + \beta_\mathfrak{o} = \#\mathfrak{o}$.

Theorem 3.2 (Shimura–Taniyama Formula [14, Section 5]). *The only slope of the Newton polygon $\nu(\mathcal{J}[\mathfrak{p}_0^\infty])$ is $\alpha_\mathfrak{o}/\#\mathfrak{o}$.*

Proof. For completeness, we briefly sketch Tate's local proof as in [14, Section 5]. First, we recall the notion of a p-divisible group with complex multiplication.

Let G be a p-divisible group defined over the ring of integers $\mathcal{O}_L$ of a finite extension L of $\mathbb{Q}_p$ such that $L \subset \mathbb{Q}_p^{\text{alg}}$. We say that G over $\mathcal{O}_L$ has complex multiplication by a local field K, for K a finite extension of $\mathbb{Q}_p$, if G has height $[K : \mathbb{Q}_p]$ and is equipped with a $\mathbb{Q}_p$-linear action of K defined over $\mathcal{O}_L$ such that, for each $\tau \in H := \text{Hom}_{\mathbb{Q}_p}(K, \mathbb{Q}_p^{\text{alg}})$,

$$\mathfrak{f}(\tau) := \dim_{\mathbb{Q}_p^{\text{alg}}}(\text{Lie}(G) \otimes_{\mathcal{O}_L} \mathbb{Q}_p^{\text{alg}})_\tau \in \{0, 1\}.$$

For $\Phi := \{\tau \in H \mid \mathfrak{f}(\tau) = 1\}$, the pair (K, Φ) is called the CM-type of G.

Let k_L denote the residue field of L and set $G_0 := G \times_{\mathcal{O}_L} k_L$, the reduction of G over k_L. We observe that, by definition, if G over $\mathcal{O}_L$ has CM-type (K, Φ), then G_0 is isoclinic (i.e., the Newton polygon of G_0 has only one slope) of slope $\#\Phi/[K : \mathbb{Q}_p]$.

Indeed, the existence of a $\mathbb{Q}_p$-linear embedding of K into $\text{End}(G) \otimes \mathbb{Q}$, with $[K : \mathbb{Q}_p] = \text{height}(G)$, implies that G_0 is isoclinic. Also, by the definition of CM-type, the dimension of G is equal to $\#\Phi$, because

$$\dim(G) = \text{rk}_{\mathcal{O}_L} \text{Lie}(G) = \sum_{\tau \in H} \mathfrak{f}(\tau) = \#\Phi.$$

We deduce that the slope of G_0 is $\frac{\dim(G)}{\text{height}(G)} = \frac{\#\Phi}{[K:\mathbb{Q}_p]}$.

To conclude, it suffices to observe that for $p \nmid m$, Lemma 3.1 implies that, for each $\mathfrak{o} \in \mathfrak{O}'$, the p-divisible group $\mathcal{J}[\mathfrak{p}_\mathfrak{o}]$, after passing to a finite extension of $\mathbb{Z}_p$,[1] has complex multiplication by $K_{d_\mathfrak{o}, \mathfrak{p}_\mathfrak{o}}$ with CM-type $(K_{d_\mathfrak{o}, \mathfrak{p}_\mathfrak{o}}, \mathfrak{o} \cap S_1)$. $\qquad\qquad\square$

---

[1] The Newton polygon is independent of the definition field of $\mathcal{J}$ and we pass to a finite extension of $\mathbb{Z}_p$ such that the CM-action is defined over this larger local ring so that we can apply Tate's theory.

Corollary 3.3. *Assume that all orbits* $\mathfrak{o} \in \mathfrak{O}'$ *are self-dual, i.e.,* $\mathfrak{o} = \mathfrak{o}^*$. *Then* $\mathcal{J}$ *has supersingular reduction at* p.

Proof. For each $\mathfrak{o} \in \mathfrak{O}'$, if $\tau \in \mathfrak{o}$, then $\tau^* \in \mathfrak{o}$. Hence $\alpha_\mathfrak{o} = \beta_\mathfrak{o}$, and the only slope of the Newton polygon $\nu(\mathcal{J}[\mathfrak{p}_\mathfrak{o}^\infty])$ is $\alpha_\mathfrak{o}/(\alpha_\mathfrak{o} + \beta_\mathfrak{o}) = 1/2$. $\square$

Remark 3.4. Let K_m^+ be the maximal totally real subfield of K_m. If each (or, equivalently, one) prime of K_m^+ above p is inert in K_m/K_m^+, then all σ-orbits $\mathfrak{o} \in \mathfrak{O}'$ are self-dual. For example, if $p \equiv -1 \bmod m$, then for all $n \in (\mathbb{Z}/m\mathbb{Z})^*$, the associated orbit is $\mathfrak{o}_n = n\langle p \rangle = \{n, -n\} = \mathfrak{o}_{-n} = \mathfrak{o}_n^*$.

Example 3.5. Let $g \geq 1$, $m = 2g + 1$, and $a = (1, 1, m - 2)$. The equation $y^m = x(x - 1)$ defines a smooth projective curve $\mathcal{C}$ over $\mathbb{Z}[1/m]$ with geometrically irreducible fibers. It has genus g by (2.2). Its Jacobian $\mathcal{J}$ over $\mathbb{Z}[1/m]$ has complex multiplication by $\prod_{1 < d|m} K_d$. Suppose that $p \nmid m$ and the order of $p \in (\mathbb{Z}/m\mathbb{Z})^*$ is even (i.e., the inertia degree of $(K_m)_\mathfrak{p}$ is even for any $\mathfrak{p}$ over p). When m is prime, this implies that all primes of K_m^+ above p are inert in K_m/K_m^+. Then $\mathcal{J}$ has supersingular reduction at p by Corollary 3.3. When $p \nmid 2m$, this is [7, Theorem 1]. See also [6, Lemma 1.1].

Remark 3.6. The Newton polygon of the Fermat curve $F_m : x^m + y^m = z^m$ is studied in [18], with the connection to Jacobi sums going back to [16]. Fix k with $2 \leq k \leq m - 1$ and consider the inertia type $a = (1, k - 1, m - k)$. Then the μ_m-Galois cover $\mathcal{C}_{m,a} \to \mathbb{P}^1$ with inertia type a is a quotient of the Fermat curve. In certain cases, this is sufficient to determine the Newton polygon of $\mathcal{C}_{m,a}$. Let f be the order of p modulo m. If f is even and $p^{f/2} \equiv -1 \bmod m$, then F_m is supersingular, so $\mathcal{C}_{m,a}$ is supersingular as well. If f is odd, then the slope $1/2$ does not occur in the Newton polygon of F_m or $\mathcal{C}_{m,a}$. Information about the p-rank of F_m and $\mathcal{C}_{m,a}$ can be found in [5].

3.2 Slopes with Large Denominators

In this section, we use Theorem 3.2 to construct curves of genus $g \geq 1$ whose Newton polygon contains only slopes with large denominators.

Consider a monodromy datum (m, a) with $m = 2g + 1$ and $a = (a(1), a(2), a(3))$.

In this section, we assume that $d \nmid a(i)$ for any $1 < d|m$.

For convenience, via the identification $\mathcal{T} = \mathbb{Z}/m\mathbb{Z}$, we identify the sets S_0 and S_1 from (3.1) as subsets of $\mathbb{Z}/m\mathbb{Z}$.

If p is a rational prime, not dividing m, we write $\langle p \rangle \subset (\mathbb{Z}/m\mathbb{Z})^*$ for the cyclic subgroup generated by the congruence class of p mod m. Then $\langle p \rangle$ acts naturally on $\mathbb{Z}/m\mathbb{Z}$. Let $n\langle p \rangle$ denote the $\langle p \rangle$-orbits in $\mathbb{Z}/m\mathbb{Z}$ where $n \not\equiv 0$.

The following proposition is a special case of Theorem 3.2.

Proposition 3.7. *Assume* $p \nmid m$. *Then the slopes of the Newton polygon* $\nu(\mathcal{J})$ *at* p *are naturally indexed by the cosets of* $\langle p \rangle$ *in* $(\mathbb{Z}/d\mathbb{Z})^*$ *for all* $1 < d \mid m$. *For each orbit* $n\langle p \rangle$, *the associated slope is*

$$\lambda_{n\langle p \rangle} := \frac{\# n \langle p \rangle \cap S_1}{\# n \langle p \rangle}.$$

Note that when the inertia type a is fixed, the Newton polygon $\nu(\mathcal{J})$ at a prime p depends only on the associated subgroup $\langle p \rangle \subset (\mathbb{Z}/m\mathbb{Z})^*$. In particular, if $m = 2g + 1$ is prime, then $\nu(\mathcal{J})$ at p depends only on the order of p in $(\mathbb{Z}/m\mathbb{Z})^*$. See also [7, Theorem 2] for the case when $a = (1, 1, m - 2)$.

Corollary 3.8. *Assume that* $m = 2g + 1$ *is prime and let* f *be a prime divisor of* g. *If* p *is a prime with* $p \nmid m$ *such that the reduction of* p *has order* f *in* $(\mathbb{Z}/m\mathbb{Z})^*$, *then every slope* λ *of the Newton polygon* $\nu(\mathcal{J})$ *at* p *with* $\lambda \neq 0, 1$ *has denominator* f. *In particular, if the* p-*rank of* $\mathcal{J}$ *is* 0, *then every slope has denominator* f.

For any m, g, and f as above, there are infinitely many primes p satisfying the hypotheses of Corollary 3.8 by the Chebotarev density theorem.

Proof. Under the hypotheses, if $\lambda_{n\langle p \rangle} \neq 0, 1$, then f is the denominator of the fraction $\lambda_{n\langle p \rangle}$ by Proposition 3.7. When the p-rank is 0, then there is no slope 0 or 1 by definition. □

Example 3.9. Let $g = 14$ and $m = 29$. For $p \equiv 7, 16, 20, 23, 24, 25 \mod 29$, the inertia degree is $f = 7$. For each choice of the inertia type, the Newton polygon is $(2/7, 5/7) \oplus (3/7, 4/7)$.

A prime number ℓ is a *Sophie Germain prime* if $2\ell + 1$ is also prime. The rest of the section focuses on the case when g is a Sophie Germain prime.

Corollary 3.10. *Suppose* g *is an odd Sophie Germain prime. Let* p *be a prime,* $p \neq 2g + 1$.

Then, one of the following occurs:

(1) *If* $p \equiv 1 \mod 2g + 1$, *then* $\nu(\mathcal{J}) = ord^g$.
(2) *If* $p \equiv -1 \mod 2g + 1$, *then* $\nu(\mathcal{J}) = ss^g$.
(3) *If* p *has order* g *modulo* $2g + 1$, *then* $\nu(\mathcal{J}) = (\alpha/g, (g - \alpha)/g)$, *for* $\alpha = \#\langle p \rangle \cap S_1$.
(4) *If* p *has order* $2g$ *modulo* $2g + 1$, *then* $\nu(\mathcal{J}) = ss^g$.

Proof. Let $m = 2g + 1$. Under the Sophie Germaine assumption on g, a prime $p \neq m$ has order either 1, 2, g, or $2g$ modulo m. Cases (2) and (4) follow from Corollary 3.3 and Cases (1) and (3) follow from Proposition 3.7. □

Note that p has order g modulo $2g + 1$ if and only if p is a quadratic residue other than 1 modulo $2g + 1$. In this case, if $a = (1, 1, 2g - 1)$, then $S_1 = \{1, \ldots, g\}$ and α is the number of quadratic residues modulo $2g + 1$ in S_1.

Example 3.11. Let $g = 5$ and $m = 11$. Let $p \equiv 3, 4, 5, 9 \bmod 11$. If $a = (1, 1, 9)$, then $v(\mathcal{J}) = (1/5, 4/5)$. If $a = (1, 2, 8)$, then $v(\mathcal{J}) = (2/5, 3/5)$.

Example 3.12. Let $g = 11$ and $m = 23$. Let $p \equiv 2, 3, 4, 6, 8, 9, 12, 13, 16, 18 \bmod 23$. If $a = (1, 1, 21)$, then $v(\mathcal{J}) = (4/11, 7/11)$. If $a = (1, 4, 18)$, then $v(\mathcal{J}) = (1/11, 10/11)$.

In Examples 3.11–3.12, the listed Newton polygons are the only ones that can occur, under these conditions on p, as a varies among all possible inertia types for μ_m-Galois covers of the projective line branched at three points. We did not find other examples of curves whose Newton polygon has slopes $1/d$ and $(d - 1)/d$ using this method.

Example 3.13. Let $g = 1013$ and $m = 2027$. Suppose the congruence class of p modulo 2027 is contained in $\langle 3 \rangle$ in $(\mathbb{Z}/2027\mathbb{Z})^*$. If the inertia type is $a = (1, 1, 2025)$, then $v(\mathcal{J}) = (523/1013, 490/1013)$.

4 Tables

The following tables contain all the Newton polygons which occur for cyclic degree m covers of the projective line branched at 3 points when $3 \leq m \leq 12$. Each inertia type $a = (a_1, a_2, a_3)$ is included, up to permutation and the action of $(\mathbb{Z}/m)^*$.

The signature is computed by (2.3) and written as $(f(1), \ldots, f(m - 1))$. We denote by *ord* the Newton polygon of $G_{0,1} \oplus G_{1,0}$ which has slopes 0 and 1 with multiplicity 1 and by *ss* the Newton polygon of $G_{1,1}$ which has slope $1/2$ with multiplicity 2.

$m = 3$			
	p	1 mod 3	2 mod 3
	Prime orbits	Split	(1, 2)
a	Signature		
(1, 1, 1)	(1, 0)	*ord*	*ss*

$m = 4$			
	p	1 mod 4	3 mod 4
	Prime orbits	Split	(1, 3), (2)
a	Signature		
(1, 1, 2)	(1, 0, 0)	*ord*	*ss*

$m = 5$				
	p	1 mod 5	2, 3 mod 5	4 mod 5
	Prime orbits	Split	$(1, 2, 3, 4)$	$(1, 4), (2, 3)$
a	Signature			
$(1, 1, 3)$	$(1, 1, 0, 0)$	ord^2	ss^2	ss^2

$m = 6$			
	p	1 mod 6	5 mod 6
	Prime orbits	Split	$(1, 5), (2, 4), (3)$
a	Signature		
$(1, 1, 4)$	$(1, 1, 0, 0, 0)$	ord^2	ss^2
$(1, 2, 3)$	$(1, 0, 0, 0, 0)$	ord	ss

$m = 7$				
	p	1 mod 7	2, 4 mod 7	3, 5 mod 7	6 mod 7
	Prime orbits	Split	$(1, 2, 4), (3, 5, 6)$	$(1, 2, 3, 4, 5, 6)$	$(1, 6), (2, 5), (3, 4)$
a	Signature				
$(1,1,5)$	$(1,1,1,0,0,0)$	ord^3	$(1/3, 2/3)$	ss^3	ss^3
$(1,2,4)$	$(1,1,0,1,0,0)$	ord^3	ord^3	ss^3	ss^3

$m = 8$				
	p	1 mod 8	3 mod 8	5 mod 8	7 mod 8
	Prime orbits	Split	$(1, 3), (2, 6)$ $(5, 7), (4)$	$(1, 5), (3, 7)$ $(2), (4), (6)$	$(1, 7), (2, 6)$ $(3, 5), (4)$
a	Signature				
$(1,1,6)$	$(1,1,1,0,0,0,0)$	ord^3	$ord^2 \oplus ss$	$ord \oplus ss^2$	ss^3
$(1,2,5)$	$(1,1,0,0,1,0,0)$	ord^3	ss^3	ord^3	ss^3
$(1,3,4)$	$(1,0,1,0,0,0,0)$	ord^2	ord^2	ss^2	ss^2

$m = 9$				
	p	1 mod 9	2, 5 mod 9	4, 7 mod 9	8 mod 9
	Prime orbits	Split	$(1, 2, 4, 8, 7, 5)$ $(3, 6)$	$(1, 4, 7), (2, 8, 5)$ $(3), (6)$	$(1, 8), (2, 7)$ $(4, 5), (3, 6)$
a	Signature				
$(1,1,7)$	$(1,1,1,1,0,0,0,0)$	ord^4	ss^4	$(1/3, 2/3) \oplus ord$	ss^4
$(1,2,6)$	$(1,1,0,0,1,0,0,0)$	ord^3	ss^3	$(1/3, 2/3)$	ss^3
$(1,3,5)$	$(1,1,0,1,0,0,0,0)$	ord^3	ss^3	$(1/3, 2/3)$	ss^3

	p	1 mod 10	3, 7 mod 10	9 mod 10
		$m = 10$		
	Prime orbits	Split	(1, 3, 9, 7), (2, 6, 8, 4), (5)	(1, 9), (2, 8), (3, 7), (4, 6), (5)
a	Signature			
(1,1,8)	(1,1,1,1,0,0,0,0,0)	ord^4	ss^4	ss^4
(1,2,7)	(1,1,1,0,0,1,0,0,0)	ord^4	ss^4	ss^4
(1,4,5)	(1,0,1,0,0,0,0,0,0)	ord^2	ss^2	ss^2

	p	1 mod 11	2, 6, 7, 8 mod 11	3, 4, 5, 9 mod 11	10 mod 11
			$m = 11$		
	Prime orbits	Split	Inert	(1, 3, 4, 5, 9), (2, 6, 7, 8, 10)	(1, 10), (2, 9), (3, 8), (4, 7), (5, 6)
a	Signature				
(1,1,9)	(1,1,1,1,1,0,0,0,0,0)	ord^5	ss^5	(1/5, 4/5)	ss^5
(1,2,8)	(1,1,1,0,0,1,1,0,0,0)	ord^5	ss^5	(2/5, 3/5)	ss^5

	p	1 mod 12	5 mod 12	7 mod 12	11 mod 12
				$m = 12$	
	Prime orbits	Split	(1, 5), (2, 10), (3), (4, 8), (6), (7, 11), (9)	(1, 7), (3, 9), (2), (4), (5, 11), (6), (8), (10)	(1, 11), (4, 8), (3, 9), (2, 10), (5, 7), (6)
a	Signature				
(1,1,10)	(1,1,1,1,1,0,0,0,0,0,0)	ord^5	$ord^3 \oplus ss^2$	$ord^2 \oplus ss^3$	ss^5
(1,2,9)	(1,1,1,0,0,0,1,0,0,0,0)	ord^4	$ord \oplus ss^3$	$ord^3 \oplus ss$	ss^4
(1,3,8)	(1,1,0,0,1,0,0,0,0,0,0)	ord^3	$ord^2 \oplus ss$	$ord \oplus ss^2$	ss^3
(1,4,7)	(1,1,0,1,0,0,1,0,0,0,0)	ord^4	ss^4	ord^4	ss^4
(1,5,6)	(1,0,1,0,1,0,0,0,0,0,0)	ord^3	ord^3	ss^3	ss^3

5 Applications

In the previous section, we computed the Newton polygons of cyclic degree m covers of the projective line branched at 3 points. We carried out the calculation of the Newton polygon for all inertia types that arise when $m \leq 23$. Many of these were not previously known to occur for the Jacobian of a smooth curve. We collect a list of the most interesting of these Newton polygons, restricting to the ones with p-rank 0 and $4 \leq g \leq 11$. In the third part of the section, we deduce some results for arbitrarily large genera g.

By [15, Theorem 2.1], if $p = 2$ and $g \in \mathbb{N}$, then there exists a supersingular curve of genus g defined over $\overline{\mathbb{F}}_2$ (or even over $\mathbb{F}_2$). For this reason, we restrict to the case that p is odd in the following result.

Theorem 5.1 (Theorem 1.1). *Let p be odd. There exists a smooth supersingular curve of genus g defined over $\overline{\mathbb{F}}_p$ in the following cases:*

Genus	Congruence	Where
4	$p \equiv 2 \bmod 3$	m=9, $a = (1, 1, 7)$
	$p \equiv 2, 3, 4 \bmod 5$	$m = 10, a = (1, 1, 8)$
5	$p \equiv 2, 6, 7, 8, 10 \bmod 11$	$m = 11$, any a
6	$p \not\equiv 1, 3, 9 \bmod 13$	$m = 13$, any a
	$p \equiv 3, 5, 6 \bmod 7$	$m = 14, a = (1, 1, 12)$
7	$p \equiv 14 \bmod 15$	$m = 15, a = (1, 1, 13)$
	$p \equiv 15 \bmod 16$	$m = 16, a = (1, 1, 14)$
8	$p \not\equiv 1 \bmod 17$	$m = 17$, any a
9	$p \equiv 2, 3, 8, 10, 12, 13, 14, 15, 18 \bmod 19$	$m = 19$, any a
10	$p \equiv 5, 17, 20 \bmod 21$	$m = 21, a = (1, 1, 19)$
11	$p \equiv 5, 7, 10, 11, 14, 15, 17, 19, 20, 21, 22 \bmod 23$	$m = 23$, any a

Proof. We compute the table using Corollary 3.3, and Remark 3.4. The genus is determined from (2.2).

For example, for $g = 5$, the congruence classes of p mod 11 are the quadratic non-residues modulo 11. A prime above p is inert in K_{11}/K_{11}^+ if and only if p is a quadratic non-residue modulo 11. (The same holds for $g = 9$ and $m = 19$, and also $g = 11$ and $m = 23$).

For example, for $g = 4$, the condition $p \equiv 3, 7, 9 \bmod 10$ covers all the cases $p \equiv 2, 3, 4 \bmod 5$ since p is odd. For $p \equiv 3, 7 \bmod 10$, p is inert in K_5. For $p \equiv -1 \bmod 10$, each orbit $\mathfrak{o} \in \mathfrak{O}'$ is self-dual. $\qquad\square$

Remark 5.2. The existence of a smooth supersingular curve of genus 9, 10, or 11 is especially interesting for the following reason. The dimension of $\mathcal{A}_g$ is $(g+1)g/2$ and the dimension of the supersingular locus in $\mathcal{A}_g$ is $\lfloor g^2/4 \rfloor$. Thus the supersingular locus has codimension 25 in $\mathcal{A}_9$, 30 in $\mathcal{A}_{10}$, and 36 in $\mathcal{A}_{11}$. The dimension of $\mathcal{M}_g$ is $3g - 3$ for $g \geq 2$. Since $\dim(\mathcal{M}_9) = 24$, $\dim(\mathcal{M}_{10}) = 27$, and $\dim(\mathcal{M}_{11}) = 30$ the supersingular locus and open Torelli locus form an unlikely intersection in $\mathcal{A}_9$, $\mathcal{A}_{10}$, and $\mathcal{A}_{11}$. See [12, Section 5.3] for more explanation.

Remark 5.3. In future work, when $5 \leq g \leq 9$, we prove there exists a smooth supersingular curve of genus g defined over $\overline{\mathbb{F}}_p$ for sufficiently large p satisfying other congruence conditions.

In the next result, we collect some other new examples of Newton polygons of smooth curves with p-rank 0.

Theorem 5.4. *There exists a smooth curve of genus g defined over $\overline{\mathbb{F}}_p$ with the given Newton polygon of p-rank 0 in the following cases:*

Genus	Newton polygon	Congruence	Where
5	$(1/5, 4/5)$	$3, 4, 5, 9 \bmod 11$	$m = 11, a = (1, 1, 9)$
5	$(2/5, 3/5)$	$3, 4, 5, 9 \bmod 11$	$m = 11, a = (1, 2, 8)$
6	$(1/3, 2/3)^2$	$3, 9 \bmod 13$	$m = 13, a = (1, 2, 10)$
		$9, 11 \bmod 14$	$m = 13, a = (1, 1, 12)$
7	$(1/4, 3/4) \oplus ss^3$	$2, 8 \bmod 15$	$m = 15, a = (1, 1, 13)$
9	$(4/9, 5/9)$	$4, 5, 6, 9, 16, 17 \bmod 19$	$m = 19, a = (1, 2, 16)$
9	$(1/3, 2/3)^3$	$4, 5, 6, 9, 16, 17 \bmod 19$	$m = 19, a = (1, 1, 17)$
		$7, 11 \bmod 19$	$m = 19, a = (1, 2, 16)$
10	$(1/3, 2/3)^3 \oplus ss$	$p \equiv 2 \bmod 21$	$m = 21, a = (1, 1, 19)$
11	$(1/11, 10/11)$	$2, 3, 4, 6, 8, 9, 12, 13, 16, 18 \bmod 23$	$m = 23, a = (1, 4, 18)$
11	$(4/11, 7/11)$	$2, 3, 4, 6, 8, 9, 12, 13, 16, 18 \bmod 23$	$m = 23, a = (1, 1, 21)$

Proof. We compute the table using the Shimura–Taniyama method, as stated in Proposition 3.7. The genus is determined from (2.2), and the signature type from (2.3).

For example, for $m = 15$ and $a = (1, 1, 13)$, the curve $C_{(m,a)}$ has genus 7 and signature type $(1, 1, \ldots, 1, 0, 0, \ldots, 0)$. Let p be a prime such that $p \equiv 2 \bmod 15$. The congruence class of p has order 2 modulo 3 and order 4 modulo 5. Hence the prime p is inert in K_3 and in K_5, and splits as a product of two primes in K_{15}. We write $\mathfrak{p}_3$ (resp. $\mathfrak{p}_5$, and $\mathfrak{p}_{15}$, $\mathfrak{p}'_{15}$) for the primes of K_3 (resp. K_5, and K_{15}) above p.

Continuing this case, by Corollary 3.3, the Newton polygon of $\mathcal{J}[\mathfrak{p}_3^\infty]$ (resp. $\mathcal{J}[\mathfrak{p}_5^\infty]$) has slope $1/2$, with multiplicity 1 (resp. 2). On the other hand, the two orbits in $(\mathbb{Z}/15\mathbb{Z})^*$ are $\mathfrak{o} = \langle 2 \rangle = \{2, 4, 8, 1\}$ and $\mathfrak{o}' = 7\langle 2 \rangle = \{7, 14, 13, 11\}$. (In particular, $\mathfrak{o}' = \mathfrak{o}^*$, hence $\mathfrak{p}'_{15} = \mathfrak{p}^*_{15}$ in K_{15}.) Hence, $\alpha_{\mathfrak{o}} = 3$ and $\alpha_{\mathfrak{o}'} = 1$. By Theorem 3.2, the Newton polygon of $\mathcal{J}[\mathfrak{p}_{15}^\infty]$ (resp. $\mathcal{J}[\mathfrak{p}'^\infty_{15}]$) has slope $3/4$ (resp. $1/4$).

For example, for $m = 21$ and $a = (1, 1, 19)$, the curve $C_{(m,a)}$ has genus 10 and signature type $(1, 1, \ldots, 1, 0, 0, \ldots, 0)$. The congruence class of $p \equiv 2 \bmod 21$ has order 2 modulo 3 and order 3 modulo 7. Hence the prime p is inert in K_3, and splits as a product of two primes in K_7 and in K_{21}. We write $\mathfrak{p}_3$ (resp. $\mathfrak{p}_7$, $\mathfrak{p}'_7$, and $\mathfrak{p}_{21}$, $\mathfrak{p}'_{21}$) for the primes of K_3 (resp. K_7, and K_{21}) above p.

Continuing this case, by Corollary 3.3, the Newton polygon of $\mathcal{J}[\mathfrak{p}_3^\infty]$ has slope $1/2$, with multiplicity 1. The two orbits in $(\mathbb{Z}/7\mathbb{Z})^*$ are $\mathfrak{o}_7 = \langle 2 \rangle = \{2, 4, 1\}$ and $\mathfrak{o}'_7 = 3\langle 2 \rangle = \{3, 6, 5\}$. Hence, $a_{\mathfrak{o}_7} = 2$ and $a_{\mathfrak{o}'_7} = 1$. By Theorem 3.2, the Newton polygon of $\mathcal{J}[\mathfrak{p}_7^\infty]$ (resp. $\mathcal{J}[\mathfrak{p}'^\infty_7]$) has slope $2/3$ (resp. $1/3$). Similarly, the two orbits in $(\mathbb{Z}/21\mathbb{Z})^*$ are $\mathfrak{o}_{21} = \langle 2 \rangle = \{2, 4, 8, 16, 11, 1\}$ and $\mathfrak{o}'_{21} = 5\langle 2 \rangle = \{5, 10, 20, 19, 17\}$. Again $a_{\mathfrak{o}_{21}} = 4$ and $a_{\mathfrak{o}'_{21}} = 2$. Thus the Newton polygon of $\mathcal{J}[\mathfrak{p}_{21}^\infty]$ (resp. $\mathcal{J}[\mathfrak{p}'^\infty_{21}]$) has slope $4/6 = 2/3$ (resp. $2/6 = 1/3$). □

Consider the p-divisible group $G_{1,d-1} \oplus G_{d-1,1}$ with slopes $1/d$, $(d - 1)/d$.

Theorem 5.5 (Theorem 1.2). *For the following values of d and p and for all $g \geq d$, there exists a smooth curve of genus g defined over $\overline{\mathbb{F}}_p$ whose Jacobian has p-divisible group isogenous to $(G_{1,d-1} \oplus G_{d-1,1}) \oplus (G_{0,1} \oplus G_{1,0})^{g-d}$:*

(1) $d = 5$ *for all* $p \equiv 3, 4, 5, 9 \bmod 11$;
(2) $d = 11$ *for all* $p \equiv 2, 3, 4, 6, 8, 9, 12, 13, 16, 18 \bmod 23$.

Proof. By Example 3.11 (resp. Example 3.12) for $d = 5$ (resp. $d = 11$), under this congruence condition on p, there exists a smooth projective curve of genus $g = d$ defined over $\overline{\mathbb{F}}_p$ whose p-divisible group is isogenous to $G_{1,d-1} \oplus G_{d-1,1}$. Note that the Newton polygon for $G_{1,d-1} \oplus G_{d-1,1}$ is the lowest Newton polygon in dimension d with p-rank 0. Thus there is at least one component of the p-rank 0 stratum of $\mathcal{M}_d$ such that the generic geometric point of this component represents a curve whose Jacobian has p-divisible group isogenous to $G_{1,d-1} \oplus G_{d-1,1}$. The result is then immediate from [12, Corollary 6.4]. □

References

1. *Revêtements étales et groupe fondamental (SGA 1)*, Documents Mathématiques (Paris) [Mathematical Documents (Paris)], 3, Société Mathématique de France, Paris, 2003, Séminaire de géométrie algébrique du Bois Marie 1960–61. [Algebraic Geometry Seminar of Bois Marie 1960–61], Directed by A. Grothendieck, With two papers by M. Raynaud, Updated and annotated reprint of the 1971 original [Lecture Notes in Math., 224, Springer, Berlin; MR0354651 (50 #7129)]. MR 2017446 (2004g:14017)
2. Jeffrey D. Achter and Rachel Pries, *Generic Newton polygons for curves of given p-rank*, Algebraic curves and finite fields, Radon Ser. Comput. Appl. Math., vol. 16, De Gruyter, Berlin, 2014, pp. 1–21. MR 3287680
3. P. Deligne and G. D. Mostow, *Monodromy of hypergeometric functions and nonlattice integral monodromy*, Inst. Hautes Études Sci. Publ. Math. (1986), no. 63, 5–89. MR 849651
4. Carel Faber and Gerard van der Geer, *Complete subvarieties of moduli spaces and the Prym map*, J. Reine Angew. Math. **573** (2004), 117–137. MR 2084584
5. Josep González, *Hasse-Witt matrices for the Fermat curves of prime degree*, Tohoku Math. J. (2) **49** (1997), no. 2, 149–163. MR 1447179
6. Benedict H. Gross and David E. Rohrlich, *Some results on the Mordell-Weil group of the Jacobian of the Fermat curve*, Invent. Math. **44** (1978), no. 3, 201–224. MR 0491708
7. Taira Honda, *On the Jacobian variety of the algebraic curve $y^2 = 1 - x^l$ over a field of characteristic $p > 0$*, Osaka J. Math. **3** (1966), 189–194. MR 0225777
8. Ju. I. Manin, *Theory of commutative formal groups over fields of finite characteristic*, Uspehi Mat. Nauk **18** (1963), no. 6 (114), 3–90. MR 0157972 (28 #1200)
9. James Milne, Complex multiplication, course notes, available at www.jmilne.org/math/CourseNotes.
10. Ben Moonen, *Special subvarieties arising from families of cyclic covers of the projective line*, Doc. Math. **15** (2010), 793–819. MR 2735989
11. Frans Oort, *Abelian varieties isogenous to a Jacobian in problems from the Workshop on Automorphisms of Curves*, Rend. Sem. Mat. Univ. Padova **113** (2005), 129–177. MR 2168985 (2006d:14027)
12. Rachel Pries, *Current results on Newton polygons of curves*, to appear in Questions in Arithmetic Algebraic Geometry, Advanced Lectures in Mathematics, Chapter 6.

13. _____, *The p-torsion of curves with large p-rank*, Int. J. Number Theory **5** (2009), no. 6, 1103–1116. MR 2569747

14. John Tate, *Classes d'isogénie des variétés abéliennes sur un corps fini (d'après T. Honda)*, Séminaire Bourbaki. Vol. 1968/69: Exposés 347–363, Lecture Notes in Math., vol. 175, Springer, Berlin, 1971, pp. Exp. No. 352, 95–110. MR 3077121

15. Gerard van der Geer and Marcel van der Vlugt, *On the existence of supersingular curves of given genus*, J. Reine Angew. Math. **458** (1995), 53–61. MR 1310953 (95k:11084)

16. André Weil, *Numbers of solutions of equations in finite fields*, Bull. Amer. Math. Soc. **55** (1949), 497–508. MR 0029393

17. Stefan Wewers, *Construction of Hurwitz spaces*, Dissertation, 1998.

18. Noriko Yui, *On the Jacobian variety of the Fermat curve*, J. Algebra **65** (1980), no. 1, 1–35. MR 578793

Arboreal Representations for Rational Maps with Few Critical Points

Jamie Juul, Holly Krieger, Nicole Looper, Michelle Manes, Bianca Thompson, and Laura Walton

Abstract Jones conjectures the arboreal representation of a degree two rational map will have finite index in the full automorphism group of a binary rooted tree except under certain conditions. We prove a version of Jones' Conjecture for quadratic and cubic polynomials assuming the *abc*-Conjecture and Vojta's Conjecture. We also exhibit a family of degree 2 rational maps and give examples of degree 3 polynomial maps whose arboreal representations have finite index in the appropriate group of tree automorphisms.

Keywords Arithmetic dynamics · Galois representations · Arboreal Galois representations · abc conjecture · Vojta's conjecture · Quadratic rational maps · Cubic polynomials

J. Juul
Department of Mathematics, The University of British Columbia, 1984 Mathematics Road, Vancouver, BC V6T 1Z2, Canada
e-mail: jamie.l.rahr@gmail.com

H. Krieger · N. Looper
Department of Pure Mathematics and Mathematical Statistics, University of Cambridge, Wilberforce Road, Cambridge CB3 0WB, UK
e-mail: hkrieger@dpmms.cam.ac.uk; nl393@cam.ac.uk

M. Manes (✉)
Department of Mathematics, University of Hawai'i at Mānoa, 2565 McCarthy Mall Keller 401A, Honolulu, HI 96822, USA
e-mail: mmanes@math.hawaii.edu

B. Thompson
Westminster College, Salt Lake City, UT 84105, USA
e-mail: bthompson@westminstercollege.edu

L. Walton
Mathematics Department, Brown University, Providence, RI 02912, USA
e-mail: laura@math.brown.edu

© The Author(s) and The Association for Women in Mathematics 2019
J. S. Balakrishnan et al. (eds.), *Research Directions in Number Theory*, Association for Women in Mathematics Series 19, https://doi.org/10.1007/978-3-030-19478-9_6

1 Introduction

Let K be a field, and fix an algebraic closure $\bar{K}$. Given a rational function $f \in K(x)$ of degree $d \geq 2$, we use f^n to denote the n-th iterate $f \circ f \circ \cdots \circ f$, and define $f^0(z) := z$.

We say $\alpha \in \mathbb{P}^1(K)$ is *periodic* if $f^n(\alpha) = \alpha$ for some $n \geq 1$; the smallest such n is called the *exact period of* α. The point α is *preperiodic* if some iterate $f^m(\alpha)$ is periodic. If all critical points of f are preperiodic, then we say the map is *post-critically finite*, or *PCF*.

Let K^s be the separable closure of K in $\bar{K}$. Choose $\alpha \in K$; for the rest of this paper we make the mild assumption that, for every $n \geq 0$, the d^n solutions to $f^n(z) = \alpha$ are distinct, thereby ensuring that these solutions live in the separable closure K^s of K.

Of recent interest, as in [3, 11, 14, 17], is the set of iterated preimages of $\alpha \in K$ under the map f:

$$\{a \in \mathbb{P}^1(K^s) : f^n(a) = \alpha \text{ for some } n \geq 0\}.$$

We consider the tree whose vertices are given by the disjoint union of the solutions to the equations

$$f^n(z) = \alpha \text{ for } n \geq 0;$$

thus, the vertex set of this tree is

$$\bigsqcup_{n \geq 0} \left\{ a \in \mathbb{P}^1(K^s) : f^n(a) = \alpha \right\}.$$

The edge relation of the tree is given by the action of f; that is, we have an edge from β_1 to β_2 if $f(\beta_1) = \beta_2$. Given the assumption above, this tree of preimages is isomorphic to the *infinite rooted d-ary tree*, which is denoted by T_∞. The tree T_∞ is sometimes denoted $T_\infty(f, \alpha)$ to show how it depends on the base point α and the function f; we suppress these in our notation. See Figure 1.

Fig. 1 The first few levels of T_∞ when pulling back $\alpha = 0$ by the polynomial $f(z) = z^2 - 2$.

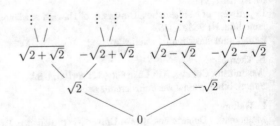

Let $\mathrm{Aut}(T_\infty)$ denote the group of automorphisms of T_∞, considered as an infinite d-ary rooted tree; this group is an infinite wreath product of S_d, the symmetric group on d letters. The absolute Galois group of K acts on T_∞ as tree automorphisms, which defines a continuous homomorphism

$$\rho_\infty \colon \mathrm{Gal}(K^s/K) \to \mathrm{Aut}(T_\infty).$$

A continuous homomorphism $\mathrm{Gal}(K^s/K) \to \mathrm{Aut}(T_\infty)$ is called an *arboreal Galois representation* [3, Definition 1.1]; the particular representation ρ_∞ defined above is called the *arboreal Galois representation associated to the pair (f, α) over K*. The study of arboreal Galois representations dates back to work of R. W. K. Odoni in the 1980s [21–23]. The image of ρ_∞, which we denote by $G_\infty(f, \alpha)$, or $G(f)$ if $\alpha = 0$, is well-studied, particularly in the degree two case [8, 9, 13, 16], and is the focus of this paper.

Jones conjectures for degree 2 rational maps that the following is true [13].

Conjecture 1. *Let K be a global field not of characteristic 2 and suppose that $f \in K(x)$ has degree two. Then $[\mathrm{Aut}(T_\infty) : G_\infty(f, \alpha)] = \infty$ if and only if one of the following holds:*

(i) *The map f is post-critically finite.*
(ii) *The two critical points γ_1 and γ_2 of f have a relation of the form $f^{r+1}(\gamma_1) = f^{r+1}(\gamma_2)$ for some $r \geq 1$.*
(iii) *The root α of T_∞ is periodic under f.*
(iv) *There is a non-trivial Möbius transformation that commutes with f and fixes α.*

The "if" direction of this conjecture is already established [13]. We will prove that assuming Vojta's Conjecture (see Conjecture 3.4.3 of [27]) for blowups of $\mathbb{P}^1 \times \mathbb{P}^1$ and the *abc*-Conjecture (see Conjecture 9), a similar set of conditions characterizes the set of quadratic and cubic polynomials $f \in K[x]$ such that $[\mathrm{Aut}(T_\infty) : G(f)] < \infty$, where K is a number field. Theorem 4 is implied by a recent result of Bridy and Tucker [4], which proves the same result when K is a number field or a function field of transcendence degree 1 over $\mathbb{Q}$. Our work, which was carried out independently, similarly makes key use of Theorem 2.10 of [10]. Our focus on the number field case allows for a more streamlined proof.

Definition 2. *Let K be a number field. We say $f \in K[x]$ is **eventually stable** if the number of irreducible factors of $f^n(x)$ over K is bounded as $n \to \infty$.*

Theorem 3. *Assume the abc-Conjecture for number fields. Let K be a number field, and let $f \in K[x]$ have degree 2. Then $[\mathrm{Aut}(T_\infty) : G(f)] = \infty$ if and only if one of the following holds:*

(i) *f is PCF*
(ii) *$f(x)$ is not eventually stable.*

Theorem 4. *Assume the abc-Conjecture for number fields, and assume Vojta's Conjecture. Let* K *be a number field, and let* $f \in K[x]$ *have degree* 3. *Then* $[\mathrm{Aut}(T_\infty) : G(f)] = \infty$ *if and only if one of the following holds:*

(i) *f is PCF*
(ii) *f(x) is not eventually stable*
(iii) *The finite critical points* γ_1, γ_2 *of* f *have a relation of the form* $f^r(\gamma_1) = f^r(\gamma_2)$ *for some* r.

Remark 1. Condition (iii) includes the case $\gamma_1 = \gamma_2$.

We use the eventual stability condition in the above theorems in place of Jones' condition on the periodicity of the root. Conjecturally, these conditions are equivalent for number fields K [15, Conjecture 1.2].

In order to prove Theorems 3 and 4 we first prove a set of sufficient conditions for finite index, Theorem 8. We then use the abc-Conjecture and Vojta's Conjecture to prove that these conditions are met. Another key ingredient in the degree 3 polynomial case is a result of Huang (also depending on Vojta's Conjecture) restricting common divisors in distinct orbits [10]. These arguments will not apply in the context of Jones' original conjecture of degree 2 rational functions since the relevant results in [10] only apply to polynomials.

The sufficient conditions of Theorem 8 and related results can be used to find examples of cubic polynomials and quadratic rational functions with finite index. In addition to providing known examples of cubic polynomials with this property, we give a new family of degree 2 rational maps and prove that the Galois groups have finite index for several parameters (in fact, they will have index 1).

Theorem 5. *Consider the family*

$$f_b(z) = \frac{z^2 - 2bz + 1}{(-2 + 2b)z}.$$

For parameters $b \in \mathbb{Z}$ *satisfying* $b \equiv 2 \mod 4$ *and* $b > 0$ *or* $b \equiv 4 \mod 8$,

$$[\mathrm{Aut}(T_\infty) : G(f_b)] = 1$$

when $K = \mathbb{Q}$ *and hence*

$$[\mathrm{Aut}(T_\infty) : G(f_b)] < \infty$$

when K *is any number field.*

2 Toward a Serre-Type Open Image Theorem for Arboreal Representations

We begin by proving a set of sufficient conditions for $[\mathrm{Aut}(T_\infty) : G(f)] < \infty$. We will use Capelli's Lemma and Dedekind's Discriminant Theorem in the proof.

Lemma 6 (Capelli's Lemma). *Let K be a field, and $f(x), g(x) \in K[x]$. Let $\alpha \in \bar{K}$ be a root of $g(x)$. Then $g(f(x))$ is irreducible over K if and only if both g is irreducible over K and $f(x) - \alpha$ is irreducible over $K(\alpha)$.*

Theorem 7 ([18], p. 100). *Let A be a Dedekind domain with field of fractions K, let L be a separable extension of K, and let B be the integral closure of A in L. If $\mathfrak{p}$ is a prime of A with $\mathfrak{p}B = \prod_i \mathfrak{q}_i^{e_i}$, and $f_i = f(\mathfrak{q}_i|\mathfrak{p})$ are the respective inertial degrees, then $\mathfrak{p}$ divides the discriminant ideal of B/A to multiplicity at least $\sum_i (e_i - 1)f_i$, with equality if, for all i, the residue characteristic of $\mathfrak{p}$ does not divide e_i.*

Theorem 8. *Let K be a number field, and let $f \in K[x]$ be a monic polynomial of degree $d \geq 2$, where d is prime, and the finite critical points of f lie in K. Suppose that $f^n(x)$ has precisely r monic irreducible factors over K for all n sufficiently large, so that*

$$f^{N+n}(x) = f_{N,1}(f^n(x))f_{N,2}(f^n(x)) \cdots f_{N,r}(f^n(x))$$

is the prime factorization of $f^{N+n}(x)$ into monic irreducibles in $K[x]$. Suppose that there is an M such that the following holds: for each $n \geq M$, and for each $1 \leq j \leq r$, there is a multiplicity one critical point γ_1 of f and a prime $\mathfrak{p}_n$ of K such that:

- $v_{\mathfrak{p}_n}(f_{N,j}(f^n(\gamma_1))) = 1$
- $v_{\mathfrak{p}_n}(f^m(\gamma_1)) = 0$ *for all $m < n + N$*
- $v_{\mathfrak{p}_n}(f^m(\gamma_t)) = 0$ *for all critical points γ_t of f with $\gamma_t \neq \gamma_1$ and all $m \leq n + N$*
- $v_{\mathfrak{p}_n}(d) = 0$ *and $v_{\mathfrak{p}_n}(a_i) \geq 0$ for all coefficients a_i of f.*

Then $[\mathrm{Aut}(T_\infty) : G(f)] < \infty$.

Remark 2. Theorem 8 does not apply to unicritical polynomials of degree at least 3, as the critical point γ_1 is required to be of multiplicity one. When f is unicritical with $d \geq 3$, it is easy to see that $[\mathrm{Aut}(T_\infty) : G(f)] = \infty$. In this case, if K contains the d-th roots of unity, then $G(f)$ is isomorphic to a subgroup of the infinite iterated wreath product of cyclic groups of order d. The geometric version of this statement follows from [17, Lemma 3.3]. The fact that we get a subgroup of this for any specialization follows from [21, Lemma 2.4].

We will make use of the following discriminant formulas from [1]. Let $\psi \in K[x]$ be of degree d with leading coefficient ℓ.

$$\mathrm{Disc}_x(\psi(x) - t) = (-1)^{(d-1)(d-2)/2} d^d \ell^{d-1} \prod_{b \in R_\psi} (t - \psi(b))^{e(b,\psi)} \tag{1}$$

where R_ψ denotes the set of critical points of ψ, $e(b, \psi)$ denotes the multiplicity of the critical point b, and t is in K. From this we obtain

$$\mathrm{Disc}_x(\psi^n(x) - t) = (-1)^{(d^n-1)(d^n-2)/2} d^{nd^n} \ell^{(d^n-1)/(d-1)} \prod_{c \in R_{\psi^n}} (t - \psi^n(c))^{e(c,\psi^n)}$$

which is equal to

$$(-1)^{(d^n-1)(d^n-2)/2} d^{nd^n} \ell^{(d^n-1)/(d-1)} \prod_{b \in R_\psi} \prod_{i=1}^{n} (t - \psi^i(b))^{e(b,\psi)}. \tag{2}$$

Proof of Theorem 8. For all $n \geq N$, let $S_{n,j}$ be the set of roots of $f^{N+n}(x)$ whose defining polynomial over K is $f_{N,j}(f^n(x))$. Let $\alpha_i \in S_{n-1,j}$. Then

$$N_{K(\alpha_i)/K}(f(\gamma_1) - \alpha_i) = f_{N,j}(f^{n-1}(f(\gamma_1))) = f_{N,j}(f^n(\gamma_1)).$$

We thus see from (1) that if for all $l \neq 1$ we have

$$v_{\mathfrak{p}_n}(d) = 0, \quad v_{\mathfrak{p}_n}(f_{N,j}(f^n(\gamma_1))) = 1, \quad v_{\mathfrak{p}_n}(f_{N,j}(f^n(\gamma_l))) = 0,$$

then $v_\mathfrak{p}(\mathrm{Disc}(f(x) - \alpha_i)) = 1$ for some prime $\mathfrak{p}$ of $K(\alpha_i)$ lying above $\mathfrak{p}_n$.

By Lemma 6, $f(x) - \alpha_i$ is irreducible over $K(\alpha_i)$, so $f(x) - \alpha_i$ is the defining polynomial of $K(\beta)/K(\alpha_i)$, where β is some preimage of α_i under f. Let S be the set of primes such that $v_\mathfrak{p}(a_i) < 0$ for some coefficient a_i of f. Then $f(x) - \alpha_i \in \mathcal{O}_{K(\alpha_i),S}[x]$. We have the relation

$$\Delta(\mathcal{O}_{K(\alpha_i),S}[\beta]/\mathcal{O}_{K(\alpha_i),S}) = [\mathcal{O}_{K(\beta),S} : \mathcal{O}_{K(\alpha_i),S}[\beta]]^2 \Delta(\mathcal{O}_{K(\beta),S}/\mathcal{O}_{K(\alpha_i),S}) \tag{3}$$

(Proposition I.3.4 of [5]). Here $\Delta(\mathcal{O}_{K(\alpha_i),S}[\beta]/\mathcal{O}_{K(\alpha_i),S})$ denotes the discriminant ideal of $\mathcal{O}_{K(\alpha_i),S}[\beta]/\mathcal{O}_{K(\alpha_i),S}$, and similarly for $\mathcal{O}_{K(\beta),S}/\mathcal{O}_{K(\alpha_i),S}$; the notation $[\mathcal{O}_{K(\beta),S} : \mathcal{O}_{K(\alpha_i),S}[\beta]]$ indicates the module index ideal of $\mathcal{O}_{K(\alpha_i),S}[\beta] \subseteq \mathcal{O}_{K(\beta),S}$. From (3), along with the fourth bulleted condition in Theorem 8 and the fact that $\Delta(\mathcal{O}_{K(\beta),S}/\mathcal{O}_{K(\alpha_i),S})$ is an integral ideal of $\mathcal{O}_{K(\alpha_i),S}$, we have

$$v_\mathfrak{p}(\Delta(\mathcal{O}_{K(\alpha_i),S}[\beta]/\mathcal{O}_{K(\alpha_i),S})) = v_\mathfrak{p}(\Delta(\mathcal{O}_{K(\beta),S}/\mathcal{O}_{K(\alpha_i),S})) = 1,$$

and

$$v_\mathfrak{p}([\mathcal{O}_{K(\beta),S} : \mathcal{O}_{K(\alpha_i),S}[\beta]]) = 0. \tag{4}$$

From Theorem 7 applied to $\mathcal{O}_{K(\beta),S}$, it follows that $e(\mathfrak{q}|\mathfrak{p}) = 2$ for any prime $\mathfrak{q}$ of the Galois closure $M_i = K(f^{-1}(\alpha_i))$ of $K(\beta)/K(\alpha_i)$ lying above $\mathfrak{p}$. Furthermore, (4) allows us to apply the Dedekind-Kummer Theorem (Proposition 8.3 of [20]) to deduce

$$f(x) - \alpha_i = (x - \eta)^2 \psi_1(x) \cdots \psi_s(x) \mod \mathfrak{p},$$

where the $\psi_i(x)$ are pairwise coprime, irreducible (and hence separable) polynomials in $\mathcal{O}_{K(\alpha_i),S}/\mathfrak{p}\mathcal{O}_{K(\alpha_i),S}$ not divisible by $x - \eta$. We conclude that $I(\mathfrak{q}|\mathfrak{p})$ acts as a transposition on the roots of $f(x) - \alpha_i$. By a standard theorem attributed to Jordan [12], if G is a primitive permutation group which is a subgroup of S_d and G contains a transposition, then $G = S_d$. Since d is prime any transitive subgroup of S_d is primitive. It follows that $\mathrm{Gal}(M_i/K(\alpha_i)) \cong S_d$.

Let $\mathfrak{Q}$ be any prime of $K_{N+n-1}M_i$ lying above $\mathfrak{q}$, and let $\mathfrak{P} = \mathfrak{Q} \cap K_{N+n-1}$. Our hypotheses on $\mathfrak{p}_n$, along with (2), imply that $\mathfrak{p}_n$ does not ramify in K_{N+n-1}. We thus have $e(\mathfrak{P}|\mathfrak{p}) = 1$, which forces $e(\mathfrak{Q}|\mathfrak{q}) = 1$ as well. Hence $e(\mathfrak{Q}|\mathfrak{P}) = 2$. Since any non-trivial element of $I(\mathfrak{Q}|\mathfrak{P})$ descends to a non-trivial element of $I(\mathfrak{q}|\mathfrak{p})$ with the same action on the roots of $f(x) - \alpha_i$, the non-trivial element of $I(\mathfrak{Q}|\mathfrak{P})$ must act as a transposition on these roots. As K_{N+n-1} is a Galois extension of $K(\alpha_i)$, $\mathrm{Gal}(K_{N+n-1}M_i/K_{N+n-1})$ is a normal subgroup of $\mathrm{Gal}(M_i/K(\alpha_i)) \cong S_d$. But a normal subgroup of S_d containing a transposition must be S_d, so we conclude that

$$\mathrm{Gal}(K_{N+n-1}M_i/K_{N+n-1}) \cong S_d.$$

Now let $\widehat{M_i} = K_{N+n-1} \prod_{j \neq i} M_j$. Let $\mathfrak{Q}'$ be a prime of K_n lying above $\mathfrak{Q}$, and let $\mathfrak{P}'$ be the prime of $\widehat{M_i}$ lying below $\mathfrak{Q}'$. We know that $\mathfrak{P}$ cannot divide $\mathrm{Disc}(f(x) - \alpha_k)$ for any root α_k of $f^{N+n-1}(x)$ with $k \neq i$; otherwise, $\mathfrak{P}$ divides either $f(\gamma_1) - \alpha_k$ or $f(\gamma_t) - \alpha_k$ for some $t \neq 1$. In the former case, $\mathfrak{P}$ then divides $\alpha_i - \alpha_k \mid \mathrm{Disc}(f^{N+n-1})$, contradicting the hypotheses on $\mathfrak{p}_n$. In the latter case, $\mathfrak{P}$ divides

$$N_{K(\alpha_j)/K}(f(\gamma_t) - \alpha_k) \mid f^{n+N}(\gamma_t),$$

also contradicting our hypotheses on $\mathfrak{p}_n$. Therefore $e(\mathfrak{P}'|\mathfrak{P}) = 1$. This forces $e(\mathfrak{Q}'|\mathfrak{Q}) = 1$, and so $e(\mathfrak{Q}'|\mathfrak{P}') = 2$.

By a similar argument to the one above, we conclude that $\mathrm{Gal}(K_{N+n}/\widehat{M_i})$ contains a transposition. As it is a normal subgroup of $\mathrm{Gal}(K_{N+n-1}M_i/K_{N+n-1}) \cong S_d$, we obtain

$$\mathrm{Gal}(K_{n+N}/\widehat{M_i}) \cong S_d,$$

and thus $\mathrm{Gal}(K_{n+N}/K_{N+n-1}) \cong S_d^m$, where $m = \deg(f^{N+n-1}) = d^{N+n-1}$. Since this holds for all sufficiently large n, $[\mathrm{Aut}(T_\infty) : G(f)] < \infty$. $\qquad\square$

We will also make use of the *abc*-Conjecture in the proofs of Theorem 3 and Theorem 4. For $(z_1, \ldots, z_n) \in K^n \setminus \{(0, \ldots, 0)\}$ with $n \geq 2$, let $N_{\mathfrak{p}} = \frac{\log(\#k_{\mathfrak{p}})}{[K:\mathbb{Q}]}$, where $k_{\mathfrak{p}}$ is the residue field of $\mathfrak{p}$, we define the *height* of the n-tuple $(z_1, \ldots, z_n)$ by

$$h(z_1, \ldots, z_n) = \sum_{\text{primes } \mathfrak{p} \text{ of } \mathcal{O}_K} -\min\{v_{\mathfrak{p}}(z_1), \ldots, v_{\mathfrak{p}}(z_n)\} N_{\mathfrak{p}}$$

$$+ \frac{1}{[K:\mathbb{Q}]} \sum_{\sigma: K \hookrightarrow \mathbb{C}} \max\{\log|\sigma(z_1)|, \ldots, \log|\sigma(z_n)|\}.$$

For any $(z_1, \ldots, z_n) \in (K^*)^n$, $n \geq 2$, we define

$$I(z_1, \ldots, z_n) = \{\text{primes } \mathfrak{p} \text{ of } \mathcal{O}_K \mid v_{\mathfrak{p}}(z_i) \neq v_{\mathfrak{p}}(z_j) \text{ for some } 1 \leq i, j \leq n\}$$

and let

$$\text{rad}(z_1, \ldots, z_n) = \sum_{\mathfrak{p} \in I(z_1, \ldots, z_n)} N_{\mathfrak{p}}.$$

Conjecture 9 (*abc*-Conjecture for Number Fields). *Let K be a number field. For any $\epsilon > 0$, there exists a constant $C_{K,\epsilon} > 0$ such that for all $a, b, c \in K^*$ satisfying $a + b = c$, we have*

$$h(a, b, c) < (1 + \epsilon)(\text{rad}(a, b, c)) + C_{K,\epsilon}.$$

Proposition 10. *Let K be a number field, and assume the abc-Conjecture for K. Let $F \in K[x]$ be a separable polynomial of degree $D \geq 3$. Then for every $\epsilon > 0$, there is a constant C_ϵ such that for every $\gamma \in K$ and every $n \geq 1$ with $F(\gamma) \neq 0$,*

$$\sum_{v_{\mathfrak{p}}(F(\gamma)) > 0} N_{\mathfrak{p}} \geq (D - 2 - \epsilon)h(\gamma) + C_\epsilon.$$

Proof. See Proposition 3.4 in [7].   □

Proposition 11 (cf. Proposition 5.1 of [7]). *Let K be a number field, and let $f \in K[x]$ be of degree $d \geq 2$. Let $\alpha \in K$ have infinite forward orbit under f. Let Z denote the set of primes of $\mathcal{O}_K$ such that $\min(v_{\mathfrak{p}}(f^m(\alpha)), v_{\mathfrak{p}}(f^n(\alpha))) > 0$ for some $m < n$ such that $f^m(\alpha) \neq 0$. Then for any $\delta > 0$, there exists an integer N such that*

$$\sum_{\mathfrak{p} \in Z} N_{\mathfrak{p}} \leq \delta h(f^n(\alpha))$$

for all $n \geq N$ with $f^n(\alpha) \neq 0$.

Proof. We have $h(f^n(\alpha)) \leq d^n(\hat{h}_f(\alpha) + O(1))$ and $h(f^n(0)) \leq d^n(\hat{h}_f(0) + O(1))$ where $\hat{h}_f$ is the canonical height, see [26, Theorem 3.20]. If $\mathfrak{p}$ divides $f^n(\alpha)$ and $f^k(\alpha)$ for some $k < n$, then $\mathfrak{p}$ divides $f^{n-k}(0)$. Therefore, any prime divisor of $f^n(\alpha)$ that divides $f^k(\alpha)$ for some $k < n$ divides either $f^k(\alpha)$ or $f^k(0)$ for some $1 \leq k \leq \lfloor n/2 \rfloor$. This yields

$$\sum_{\mathfrak{p} \in Z} N_{\mathfrak{p}} \leq \sum_{i=1}^{\lfloor n/2 \rfloor} h(f^i(\alpha)) + h(f^i(0))$$

$$\leq \sum_{i=1}^{\lfloor n/2 \rfloor} d^i(\hat{h}_f(0) + \hat{h}_f(\alpha) + O(1))$$

$$\leq d^{\lfloor n/2 \rfloor + 1}(\hat{h}_f(0) + \hat{h}_f(\alpha) + O(1))$$

$$\leq \delta h(f^n(\alpha))$$

for all sufficiently large n. $\qquad\square$

The proof of the following proposition is similar to that of [7, Proposition 3.4].

Proposition 12. *Let $f(x) \in K[x]$ where K is a number field, and assume the abc-Conjecture for K. Suppose that f has degree $d \geq 2$ and that $f^n(x)$ has precisely r irreducible factors over K for all n sufficiently large, so that*

$$f^{N+n}(x) = f_{N,1}(f^n(x)) f_{N,2}(f^n(x)) \cdots f_{N,r}(f^n(x))$$

is the prime factorization of $f^{N+n}(x)$ into monic irreducibles in $K[x]$. Let $\gamma \in K$ have infinite forward orbit under f. Let Z denote the set of primes of $\mathcal{O}_K$ such that $\min(v_{\mathfrak{p}}(f^m(\alpha)), v_{\mathfrak{p}}(f^{n+N}(\alpha))) > 0$ for some $m < n + N$ such that $f^m(\alpha) \neq 0$. Fix $1 \leq j \leq r$. Then for all sufficiently small δ, we have

$$\sum_{\substack{\mathfrak{p} \notin Z \\ v_{\mathfrak{p}}(f_{N,j}(f^n(\gamma)))=1}} N_{\mathfrak{p}} \geq \delta h(f_{N,j}(f^n(\gamma)))$$

for all $n \gg_\delta 0$. In particular, for all $n \gg_\delta 0$, there is a prime $\mathfrak{p} \notin Z$ such that $v_{\mathfrak{p}}(f_{N,j}(f^n(\gamma))) = 1$.

Proof. Choose an i so that $f_{N,j}(f^i(x))$ has degree $D \geq 8$. By Proposition 10 with $\epsilon = 1$,

$$\sum_{v_{\mathfrak{p}}(f_{N,j}(f^n(\gamma)))>0} N_{\mathfrak{p}} \geq (D-3)h(f^{n-i}(\gamma)) + C_1. \tag{5}$$

On the other hand, since $h(\psi(z)) \leq dh(z) + O_\psi(1)$ for any rational function ψ of degree d (see [26, Theorem 3.11]), we also have

$$\sum_{v_{\mathfrak{p}}(f_{N,j}(f^n(\gamma)))>0} v_{\mathfrak{p}}(f_{N,j}(f^n(\gamma)))N_{\mathfrak{p}} \leq Dh(f^{n-i}(\gamma)) + O(1).$$

From this, we can see that

$$\sum_{v_{\mathfrak{p}}(f_{N,j}(f^n(\gamma)))\geq 2} N_{\mathfrak{p}} \leq \frac{D}{2}h(f^{n-i}(\gamma)) + O(1). \tag{6}$$

Combining (5) and (6) gives

$$\sum_{v_{\mathfrak{p}}(f_{N,j}(f^n(\gamma)))=1} N_{\mathfrak{p}} \geq \left(\frac{D}{2} - 3\right) h(f^{n-i}(\gamma)) + C_2 \geq h(f^{n-i}(\gamma)) + C_2.$$

Now by Proposition 11 with $\delta = \frac{1}{2d^{N+i}}$ we see that for all sufficiently large n,

$$\sum_{\mathfrak{p}\in Z} N_{\mathfrak{p}} < \frac{1}{2d^{N+i}}h(f^{N+n}(\gamma)) < \frac{1}{2}h(f^{n-i}(\gamma)) + C_3.$$

Hence,

$$\left(\sum_{v_{\mathfrak{p}}(f_{N,j}(f^n(\gamma)))=1} N_{\mathfrak{p}}\right) - \left(\sum_{\mathfrak{p}\in Z} N_{\mathfrak{p}}\right) > \frac{1}{2}h(f^{n-i}(\gamma)) + C_2 - C_3.$$

Since we have fixed i, we see that there exists a $\delta = \delta(i, N) > 0$ such that for all $n \gg_\delta 0$,

$$\sum_{\substack{v_{\mathfrak{p}}(f_{N,j}(f^n(\gamma)))=1 \\ \mathfrak{p}\notin Z}} N_{\mathfrak{p}} \geq \delta h(f_{N,j}(f^n(\gamma))). \qquad \square$$

We are now ready to prove Theorem 3.

Proof of Theorem 3. We can see that each of the conditions are sufficient for $[\mathrm{Aut}(T_\infty) : G(f)] = \infty$ by the comments after Conjecture 3.11 of [13].

Now suppose that f is not PCF and f is eventually stable. We can conjugate f by scaling to give a monic polynomial. Then we can see by Proposition 12 that the conditions of Theorem 8 are met. Hence, $[\mathrm{Aut}(T_\infty) : G(f)] < \infty$. $\qquad \square$

A key ingredient in our proof of the degree 3 polynomial case is the following result from [10].

For $a, b \in K^*$, define the generalized greatest common divisor of a and b as

$$h_{\text{gcd}}(a, b) = \frac{1}{[K : \mathbb{Q}]} \sum_{v \in M_K^0} n_v \min(v^+(a), v^+(b))$$

where $v^+(a) = \max\{-\log |a|_v, 0\}$, M_K^0 denotes the non-archimedean places of K, and $n_v = [K_v : \mathbb{Q}_v]$. We say a point $y \in \overline{K}$ is *exceptional for* f if $\cup_{n=0}^\infty f^{-n}(y)$ is finite.

Theorem 13 ([10, Theorem 2.10]). *Assume Vojta's Conjecture (Conjecture 3.4.3 of [27]). Let K be a number field and $f \in K[x]$ be a polynomial of degree $d \geq 2$. Assume that f is not conjugate (by a rational automorphism defined over $\overline{\mathbb{Q}}$) to a power map or a Chebyshev map. Suppose $0, a, b \in K$ are not exceptional for f. Assume that there is no polynomial $H \in \overline{\mathbb{Q}}[x]$ such that: $H \circ f^k = f^k \circ H$ for some $k \geq 1$, $H(0) = 0$, and $H(f^l(a)) = f^m(b)$ or $H(f^l(b)) = f^m(a)$ for some $l, m \geq 1$. Then for any $\epsilon > 0$, there exists a $C = C(\epsilon, a, b, f) > 0$ such that for all $n \geq 1$, we have*

$$h_{\text{gcd}}(f^n(a), f^n(b)) \leq \epsilon d^n + C.$$

Remark 3. Theorem 2.10 of [10] is in fact more general; we have simply taken $\alpha = \beta = 0$ in its statement.

Definition 14. *For $y \in K$ and $f(x) \in K[x]$, we say that a prime $\mathfrak{p}$ of K is a primitive prime divisor of $f^n(y)$ if $v_{\mathfrak{p}}(f^n(y)) > 0$ and $v_{\mathfrak{p}}(f^m(y)) = 0$ for all $1 \leq m \leq n - 1$.*

Proposition 15. *Let K be a number field, and let $f \in K[x]$ be a non-PCF, eventually stable polynomial of degree 3 with distinct finite critical points γ_1, γ_2. If $f^i(\gamma_1) = f^j(\gamma_2)$ for some $i \neq j$, then $[\text{Aut}(T_\infty) : G(f)] < \infty$.*

Proof. As f is non-PCF, the hypothesis implies that γ_1 and γ_2 have infinite forward orbit under f. If $f^i(\gamma_1) = f^j(\gamma_2)$ for $i < j$, then for all $n > 2j - i$, any prime divisor of $f^n(\gamma_2)$ by definition is not a primitive prime divisor of $f^n(\gamma_1)$. The result then follows immediately from Theorem 8 combined with Proposition 12. □

Proposition 16. *Let K be a number field, and let $f \in K[x]$ be a non-PCF polynomial of degree 3 with distinct finite critical points $\gamma_1, \gamma_2 \in K$. Suppose there exists an $H \in \overline{\mathbb{Q}}[x]$ such that $H \circ f^t = f^t \circ H$ for some $t \geq 1$, and that $H(0) = 0$, and $H(f^m(\gamma_1)) = f^n(\gamma_2)$ for some $m, n \geq 1$. Then $H = L \circ \psi^r$, $r \geq 0$, where L is a linear polynomial commuting with some iterate of f, and $\psi^k = f^l$ for some $k, l \geq 0$. One of the following must hold:*

(i) $L(x) = ax$ with $a \neq 1$, in which case f is not eventually stable
(ii) $L(x) = x$, in which case $f^i(\gamma_1) = f^j(\gamma_2)$ for some $i, j \geq 1$
(iii) $L(0) \neq 0$, in which case, for any $\epsilon > 0$, $h_{\text{gcd}}(f^n(\gamma_1), f^n(\gamma_2)) \leq \epsilon d^n$ for all sufficiently large n.

Proof. That there exist such an L and ψ follows from Ritt's Theorem [24, 25]. Case (ii) is clear. In Case (i), the fact that L commutes with an iterate of f implies that f has trivial constant term, and hence fails to be eventually stable. Thus suppose we are in Case (iii). Conjugating f to a monic centered representative g, so that $g = \mu^{-1} f \mu$, we obtain $H' = L' \circ \psi'^r$, where $H' = \mu^{-1} H \mu$, $L' = \mu^{-1} L \mu$, and $\psi' = \mu^{-1} \psi \mu$. As g is in monic centered form, we must have $L'(x) = -x$ or $L'(x) = x$. If $L'(x) = x$, then we are in Case (ii); assume therefore that $L'(x) = -x$. If γ_1' and γ_2' are the critical points of g, then this implies $g^n(\gamma_1') = -g^n(-\gamma_1') = -g^n(\gamma_2')$. Noting that $\mu^{-1}(0) \neq 0$ (for otherwise, Case (i) or Case (ii) would hold), it follows that

$$h_{\gcd}(g^n(\gamma_1') - \mu^{-1}(0), g^n(\gamma_2') - \mu^{-1}(0)) = h_{\gcd}(f^n(\gamma_1), f^n(\gamma_2)) \leq \epsilon d^n$$

for all sufficiently large n.   □

Proof of Theorem 4. The sufficiency of conditions (i) and (ii) follows from the proof of Conjecture 3.11 of [13]. To see that (iii) implies the index is infinite, note that

$$\mathrm{Disc}(f^n) = \pm 3^{3^n} \mathrm{Disc}(f^{n-1})^3 f^n(\gamma_1) f^n(\gamma_2).$$

Replacing K by a finite extension if necessary, we can assume that K contains i and $\sqrt{3}$. Then $\mathrm{Disc}(f^n)$ is a square in K_{n-1}, and so $\mathrm{Gal}(K_n/K_{n-1})$ contains only even permutations in its action on the roots of f^n. Since this holds for all $n \geq r$, $[\mathrm{Aut}(T_\infty) : G(f)] = \infty$.

Conversely, assume f is not PCF, that f is eventually stable, and that f has two distinct finite critical points $\gamma_1, \gamma_2 \in K$ not satisfying (iii) for any r. The assumption that f is not PCF means that 0, γ_1, and γ_2 are not exceptional points of f [2, §4.1]. If there is no polynomial H as described in Theorem 13, then Proposition 12 and Theorem 13 allow us to apply Theorem 8 and conclude that $[\mathrm{Aut}(T_\infty) : G(f)] < \infty$. (Note that by passing to a quadratic extension of K if needed, we can conjugate by scaling to assume f is monic, as in the statement of Theorem 8.) On the other hand, if there does exist such an H, then by Proposition 16, either $f^i(\gamma_1) = f^j(\gamma_2)$ for some $i, j \geq 1$ with $i \neq j$, or, for any $\epsilon > 0$, we have

$$h_{\gcd}(f^n(\gamma_1), f^n(\gamma_2)) \leq \epsilon d^n$$

for all sufficiently large n. In the former case, Proposition 15 then yields $[\mathrm{Aut}(T_\infty) : G(f)] < \infty$. In the latter case, Theorem 8 and Proposition 12 complete the proof.   □

3 Example of Families of Rational Maps with Finite Index

3.1 Cubic Polynomial Examples

If we focus on specific examples of cubic and quadratic families of maps, we need not appeal to the *abc*-Conjecture or Vojta's Conjecture to get finite index results. The following proposition gives sufficient conditions one can check for a specific example or family of cubic polynomials to produce finite index arboreal representations.

Proposition 17. *Let K be a number field and let $f(x) \in K[x]$ be a monic degree 3 polynomial. Let γ_1, γ_2 denote the finite critical points of f. Suppose that $f^n(x)$ is irreducible and there exists a prime $\mathfrak{p}$ of K such that*

- $v_{\mathfrak{p}}(f^n(\gamma_1))$ *is odd*
- $v_{\mathfrak{p}}(f^n(\gamma_2)) = 0$
- $v_{\mathfrak{p}}(3) = 0$
- $v_{\mathfrak{p}}(f^i(\gamma_1)) = v_{\mathfrak{p}}(f^i(\gamma_2)) = 0$ *for* $1 \leq i < n$.

Then $[K_n : K_{n-1}] = 6^{3^{n-1}}$.

Proof. Let $\alpha_1, \alpha_2, \ldots, \alpha_{3^{n-1}}$ denote the roots of $f^{n-1}(x)$. As in Section 2, let $M_i = K(f^{-1}(\alpha_i))$ and $\widehat{M_i} = K_{n-1} \prod_{i \neq j} M_j$.

Since $v_{\mathfrak{p}}(f^n(\gamma_1))$ is odd, we must have $v_{\mathfrak{p}'}(\mathrm{Disc}(f(x) - \alpha_i)$ is odd for some prime $\mathfrak{p}'$ of $K(\alpha_i)$ lying over $\mathfrak{p}$. Thus, $\mathfrak{p}'$ must ramify in M_i, and $f(x) - \alpha_i$ must have at least a double root modulo $\mathfrak{p}'$. If this root had multiplicity 3, then $f'(x)$ has a double root modulo $\mathfrak{p}'$, so $\gamma_1 \equiv \gamma_2 \mod \mathfrak{p}'$ and hence $f^n(\gamma_1) \equiv f^n(\gamma_2) \mod \mathfrak{p}'$, contradicting our hypotheses. Hence, $f(x) - \alpha_i \equiv (x - \eta)^2(x - \xi) \mod \mathfrak{p}'$ and for any prime $\mathfrak{q}$ of $K(\alpha_i)$ lying over $\mathfrak{p}'$, $I(\mathfrak{q}|\mathfrak{p}')$ acts as a transposition on the roots of $f(x) - \alpha_i$.

Following the arguments in the proof of Theorem 8, we can see that $\mathrm{Gal}(K_n/\widehat{M_i}) \cong S_3$ and hence $[K_n : K_{n-1}] = 6^{3^{n-1}}$. $\qquad\square$

Consider the cubic polynomials:

$$g_1(z) = z^3 - \frac{6012}{2755}z^2 + \frac{12636}{13775}z + \frac{54}{95}$$

and

$$g_2(z) = z^3 + 7z^2 - 7.$$

In [6], Combs shows that the image of the arboreal representation associated to g_1 is surjective. In [19], Looper shows that the image of the arboreal representation associated to g_2 is an index 2 subgroup of $\mathrm{Aut}(T_\infty)$. Each of these examples can be shown to satisfy the hypotheses of Proposition 17 for all $n \geq 2$.

3.2   A Family of Degree 2 Rational Maps

In order to give an example of a family of quadratic rational map with surjective arboreal Galois representation, we follow the conventions of [16]. To this end, we define the following notation:

$f^n = \frac{P_n}{Q_n}$ where P_n and Q_n are polynomials;

$\ell(r)$ the leading coefficient of a polynomial r;

$v_p(n)$ the p-adic valuation of n for some prime p,
               i.e., if $n = p^v d$ with $p \nmid d$, then $v_p(n) = v$; and

$\text{Res}(Q, P)$ the resultant of the polynomials P and Q.

Proposition 17 is analogous to a result of Jones and Manes [16, Corollary 3.8], which we state in the following theorem.

Theorem 18 ([16, Corollary 3.8]). *Let $f = \frac{P_1(x)}{Q_1(x)} \in K(x)$ have degree 2, let $c = Q_1 P_1' - P_1 Q_1'$, and suppose that $f^n(\infty) \neq 0$ for all $n \geq 1$ and that f has two finite critical points γ_1, γ_2 with $f(\gamma_i) \neq \infty$ for each i. Suppose further that there exists a prime $\mathfrak{p}$ of K with $v_{\mathfrak{p}}(P_n(\gamma_1) P_n(\gamma_2))$ odd and*

$$0 = v_{\mathfrak{p}}(\ell(P_1)) = v_{\mathfrak{p}}(\ell(c)) = v_{\mathfrak{p}}(\text{Res}(Q_1, P_1)) = v_{\mathfrak{p}}(\text{Disc } P_1) = v_{\mathfrak{p}}(P_j(\gamma_i))$$

for $1 \leq i \leq 2, 2 \leq j \leq n - 1$. Then $[K_n : K_{n-1}] = 2^{2^{n-1}}$.

Consider the family

$$f_b(z) = \frac{z^2 - 2bz + 1}{(-2 + 2b)z},$$

where $b \neq 1$ is an algebraic number.

We will use Theorem 18 to show that an infinite number of members of this family have surjective arboreal representation over $\mathbb{Q}$. This is the first example of an infinite family of non-polynomial rational maps having finite index arboreal representation. Prior to this work, there was a single non-polynomial quadratic map that was known to have surjective arboreal representation; this example was given in [16, Theorem 1.2].

In the context of this particular family, we see that:

$$P_1(x) = z^2 - 2bz + 1,$$

$$Q_1(x) = (-2 + 2b)z,$$

$$\ell(P_1) = 1,$$

$$\text{Disc}(P_1) = 4(b^2 - 1),$$

$$c = (-2 + 2b)z(2z - 2b) - (z^2 - 2bz + 1)(-2 + 2b)$$
$$= (-2 + 2b)(z^2 - 1),$$
$$\ell(c) = 2(b - 1),$$
$$\mathrm{Res}(Q_1, P_1) = 4(b - 1)^2,$$
$$P_n(z) = P_{n-1}(z)^2 - 2bP_{n-1}(z)Q_{n-1}(z) + Q_{n-1}(z)^2,$$
$$Q_n(z) = 2(b - 1)P_{n-1}(z)Q_{n-1}(z).$$

The point ∞ is a fixed point for this family. Since every member of this family maps 0 to ∞, we see that 0 is always *strictly* preperiodic like the example done in [16, Theorem 1.2]. The critical points of this family are 1 and -1. The critical point 1 is mapped by every member of the family to the critical point -1. So the critical orbit behavior of this family is constrained by the fact that the critical orbits collide. Whereas in Theorem 1.2 of [16], the critical orbit behavior was constrained by the fact that one critical point was periodic.

As long as the numerator of $f_b(z)$ is irreducible, the extension at level $n = 1$ is maximal. Over $\mathbb{Q}$, this is the case for any integer $b \neq 0, 1$. We will show that, for all $n \geq 2$, the numerator $P_n(-1)$ has a primitive prime divisor with odd valuation that does not divide $2(b - 1)(b + 1)$.

Lemma 19. *Suppose $b \neq 1$ is an integer. Let p be a prime and suppose that $p \mid Q_i(-1)$ for some $i \geq 1$. Then, for all $j > i$, we have that $p \mid Q_j(-1)$.*

Proof. Suppose that $p \mid Q_i(-1)$. Then,

$$Q_{i+1}(-1) = 2(b - 1)P_i(-1)Q_i(-1),$$

so $p \mid Q_{i+1}(-1)$. By induction, the result follows. □

Lemma 20. *Suppose $b \neq 1$ is an integer. Then, for $n \geq 1$, the only possible common divisors of $P_n(-1)$ and $Q_n(-1)$ are powers of 2.*

Proof. We proceed by induction. For the base case, we remark that $P_1(-1) = 2 + 2b$, and $Q_1(-1) = 2 - 2b$. Thus, any common factor of $P_1(-1)$ and $Q_1(-1)$ must divide $P_1(-1) + Q_1(-1) = 4$, which gives the base case.

Suppose that the result is true for $P_n(-1)$ and $Q_n(-1)$. We wish to show that the same is true for $P_{n+1}(-1)$ and $Q_{n+1}(-1)$.

Suppose that p is an odd prime dividing $Q_{n+1}(-1)$. Since

$$Q_{n+1}(-1) = 2(b - 1)P_n(-1)Q_n(-1),$$

it follows that $p \mid (b - 1)$, or that $p \mid Q_n(-1)$, or that $p \mid P_n(-1)$. Moreover, if $p \mid (b-1)$, then $p \mid Q_n(-1)$ by Lemma 19. Thus, either $p \mid Q_n(-1)$ or $p \mid P_n(-1)$; by the inductive hypothesis, p cannot divide both.

Suppose that $p \mid Q_n(-1)$ and $p \nmid P_n(-1)$; the argument in the other case is identical. Then,

$$P_{n+1}(-1) = P_n(-1)^2 - 2bP_n(-1)Q_n(-1) + Q_n(-1)^2 \equiv P_n(-1)^2 \not\equiv 0 \mod p,$$

so $p \nmid P_{n+1}(-1)$. $\square$

Lemma 21. *Suppose $b \neq 1$ is an integer. Let p be an odd prime and suppose that $p \mid P_i(-1)$ for some $i \geq 1$. Then, for all $j > i$, we have that $p \mid Q_j(-1)$ and $p \nmid P_j(-1)$.*

Proof. Suppose p is an odd prime and $p \mid P_i(-1)$. Since

$$Q_{i+1}(-1) = 2(b-1)P_i(-1)Q_i(-1),$$

it follows that $p \mid Q_{i+1}(-1)$. By Lemma 19, it follows that $p \mid Q_j(-1)$ for all $j > i$. Since p is an odd prime, it follows by Lemma 20 that $p \nmid P_j(-1)$ for all $j > 1$. $\square$

Lemma 22. *Suppose $b \neq 1$ is an integer. If p is an odd prime divisor of $P_n(-1)$, then p is a primitive prime divisor. Furthermore, for any $n \geq 2$, the odd primes dividing $P_n(-1)$ do not divide $2(b-1)(b+1)$.*

Proof. Suppose that p is an odd prime divisor of $P_n(-1)$. Suppose further that $p \mid P_i(-1)$ for some $i < n$. By Lemma 21, it follows that $p \nmid P_j(-1)$ for all $j > i$; in particular, this implies that $p \nmid P_n(-1)$, a contradiction. Hence, any odd prime dividing $P_n(-1)$ is primitive.

Note that $P_1(-1) = 2 + 2b$, and $Q_1(-1) = 2 - 2b$. Any odd prime p dividing $2(b-1)(b+1)$ must therefore divide $P_1(-1) = 2 + 2b$ or $Q_1(-1) = 2 - 2b$. By Lemmas 19, 20, and 21, it follows that $p \nmid P_n(-1)$ for $n > 1$. $\square$

Lemma 23. *Let b be an even integer. Then $v_2(P_n(-1)) = v_2(Q_n(-1)) = 2^n - 1$.*

Proof. We proceed by induction on n. Since b is even $v_2(2+2b) = v_2(2-2b) = 1$, so the result holds for $n = 1$. Now suppose

$$v_2(P_{n-1}(-1)) = v_2(Q_{n-1}(-1)) = 2^{n-1} - 1.$$

Write $P_{n-1} = 2^{2^{n-1}-1}u_{n-1}$ and $Q_{n-1} = 2^{2^{n-1}-1}w_{n-1}$ where u_{n-1} and w_{n-1} are relatively prime odd integers. Then

$$Q_n(-1) = 2(b-1)2^{2^{n-1}-1}u_{n-1}2^{2^{n-1}-1}w_{n-1}$$
$$= 2^{2^n-1}(b-1)u_{n-1}w_{n-1}$$

and

$$P_n(-1) = (2^{2^{n-1}-1}u_{n-1})^2 - 2b(2^{2^{n-1}-1}u_{n-1})(2^{2^{n-1}-1}w_{n-1}) + (2^{2^{n-1}-1}w_{n-1})^2$$
$$= 2^{2^n-2}(u_{n-1}^2 - 2bu_{n-1}w_{n-1} + w_{n-1}^2).$$

Since u_{n-1} and w_{n-1} are odd and b is even, $u_{n-1}^2 - 2bu_{n-1}w_{n-1} + w_{n-1}^2 \equiv 2$ mod 4. Hence $v_2(P_n) = v_2(Q_n) = 2^n - 1$ as desired.   □

Lemma 24. *Let* $b \equiv 2 \mod 4$ *and* $b > 0$. *Then* $P_n(-1) = 2^{2^n-1}u_n$ *where* u_n *is odd and* $u_n \neq \pm y^2$.

Proof. We can see by induction that $P_n(-1) > 0$ and $Q_n(-1) < 0$. Thus, $u_n > 0$ so $u_n \neq -y^2$ for any integer y. By the proof of Lemma 23,

$$2u_n = u_{n-1}^2 - 2bu_{n-1}w_{n-1} + w_{n-1}^2,$$

where u_{n-1} and w_{n-1} are odd. Thus, $2u_n \equiv 6 \mod 8$. This implies $u_n \equiv 3$ or 7 mod 8 and hence $u_n \neq y^2$ for any integer y.   □

Lemma 25. *Suppose* $b \equiv 4 \mod 8$. *Then* $P_n(-1) = 2^{2^n-1}u_n$ *where* u_n *is odd and* $u_n \neq \pm y^2$.

Proof. We claim that u_n and w_n are each congruent to $\pm 3 \mod 8$. We proceed by induction. First note that $u_1 = 1 + b$ and $w_1 = 1 - b$, and since $b \equiv 4 \mod 8$, the result holds for $n = 1$.

Now suppose and u_{n-1} and w_{n-1} are both $\pm 3 \mod 8$. Then,

$$w_n = (b-1)u_{n-1}w_{n-1} \equiv \pm 3^3 \equiv \pm 11 \mod 16,$$

and

$$2u_n = u_{n-1}^2 - 2bu_{n-1}w_{n-1} + w_{n-1}^2$$
$$\equiv 9 + 8 + 9 \equiv 10 \mod 16.$$

Hence u_n must be congruent to 5 or 13 modulo 16, thus proving the claim that u_n and w_n are each congruent to $\pm 3 \mod 8$. This shows that $u_n \neq \pm y^2$ for any integer y.   □

Proof of Theorem 5. By design $f_b^n(\infty) \neq 0$ for all $n \geq 1$ and f_b has two finite critical points 1 and -1 whose orbits collide. That is, $f_b(1) = -1$. Since $f_b^n(1) = f_b^{n-1}(-1)$, we have

$$\frac{P_n(1)}{Q_n(1)} = \frac{P_{n-1}(-1)}{Q_{n-1}(-1)}.$$

Noting that $P_1(1) = 2 - 2b = -(-2 + 2b) = -Q_1(1)$, it follows by induction from the relation

$$Q_n(1) = 2(b - 1)P_{n-1}(1)Q_{n-1}(1)$$

that any common divisor of $P_n(1)$ and $Q_n(1)$ must be a common divisor of $P_1(1)$ and $Q_1(1)$, and hence must divide $2(b - 1)$. Lemmas 22, 24, and 25 then imply that each $n \geq 2$, we have a prime p such that $v_p(P_n(1)P_n(-1)) = v_p(P_n(-1))$ is odd and

$$0 = v_p(\ell(P_1)) = v_p(\ell(c)) = v_p(\text{Res}(Q_1, P_1)) = v_p(\text{Disc } P_1) = v_p(P_j(\pm 1)),$$

for all $2 \leq j \leq n - 1$. Thus we can apply Theorem 18 to conclude that $[K_n : K_{n-1}] = 2^{2^{n-1}}$ for each n, proving $[\text{Aut}(T_\infty) : G(f_b)] = 1$. □

Remark 4. These arguments do not readily extend to $b \equiv 2^n \mod 2^{n+1}$ for $n \geq 3$. We note that the proofs of Lemmas 24 and 25 yield squares in the base case when $2^n + 1 = y^2$, and the induction step implies that $u_n, w_n \equiv 1 \mod 2^{n+1}$. A different argument is needed for other values of b.

Acknowledgements This project began at the Women in Numbers 4 conference at BIRS. We would like to thank BIRS for hosting WIN4 and the AWM Advance Grant for funding the workshop. We would also like to thank an anonymous referee for very careful reading of the first draft of this article.

MM was partially supported by the Simons Foundation grant #359721. NL was partially supported by an NSF Graduate Research Fellowship.

This material is based on work supported by and while the fourth author served at the National Science Foundation. Any opinion, findings, and conclusions or recommendations expressed in this material are those of the authors and do not necessarily reflect the views of the National Science Foundation.

References

1. Wayne Aitken, Farshid Hajir, and Christian Maire. Finitely ramified iterated extensions. *Int. Math. Res. Not.*, (14):855–880, 2005.
2. Alan F. Beardon. *Iteration of rational functions*, volume 132 of *Graduate Texts in Mathematics*. Springer-Verlag, New York, 1991.
3. Nigel Boston and Rafe Jones. Arboreal Galois representations. *Geom. Dedicata*, 124:27–35, 2007.
4. Andrew Bridy and Tom Tucker. Finite index theorems for iterated galois groups of cubic polynomials. *arXiv preprint arXiv:1710.02257*, 2017.
5. J. W. S. Cassels and A. Fröhlich. Hans Arnold Heilbronn. *Bull. London Math. Soc.*, 9(2):219–232. (1 plate), 1977.
6. TJ Combs. *Arboreal Galois representations of cubic polynomials with full image*. PhD thesis, University of Hawaii at Manoa, 2018.
7. C. Gratton, K. Nguyen, and T. J. Tucker. ABC implies primitive prime divisors in arithmetic dynamics. *Bull. Lond. Math. Soc.*, 45(6):1194–1208, 2013.

8. Spencer Hamblen, Rafe Jones, and Kalyani Madhu. The density of primes in orbits of $z^d + c$. *Int. Math. Res. Not. IMRN*, (7):1924–1958, 2015.

9. Wade Hindes. Galois uniformity in quadratic dynamics over $k(t)$. *J. Number Theory*, 148:372–383, 2015.

10. Keping Huang. Generalized greatest common divisors for the orbits under rational functions. arXiv:1702.03881 [math.NT], 2017.

11. Patrick Ingram. Arboreal Galois representations and uniformization of polynomial dynamics. *Bull. Lond. Math. Soc.*, 45(2):301–308, 2013.

12. I. Martin Isaacs. *Finite group theory*, volume 92 of *Graduate Studies in Mathematics*. American Mathematical Society, Providence, RI, 2008.

13. Rafe Jones. Galois representations from pre-image trees: an arboreal survey. In *Actes de la Conférence "Théorie des Nombres et Applications"*, Publ. Math. Besançon Algèbre Théorie Nr., pages 107–136. Presses Univ. Franche-Comté, Besançon, 2013.

14. Rafe Jones and Nigel Boston. Settled polynomials over finite fields. *Proc. Amer. Math. Soc.*, 140(6):1849–1863, 2012.

15. Rafe Jones and Alon Levy. Eventually stable rational functions. *Int. J. Number Theory*, 13(9):2299–2318, 2017.

16. Rafe Jones and Michelle Manes. Galois theory of quadratic rational functions. *Comment. Math. Helv.*, 89(1):173–213, 2014.

17. Jamie Juul, Pär Kurlberg, Kalyani Madhu, and Tom J. Tucker. Wreath products and proportions of periodic points. *Int. Math. Res. Not. IMRN*, (13):3944–3969, 2016.

18. H. Koch. *Algebraic number theory*. Springer-Verlag, Berlin, Russian edition, 1997. Reprint of the 1992 translation.

19. Nicole Looper. Dynamical Galois groups of trinomials and Odoni's conjecture. arXiv:1609.03398v2 [math.NT], 2016.

20. Jürgen Neukirch. *Algebraic number theory*, volume 322 of *Grundlehren der Mathematischen Wissenschaften [Fundamental Principles of Mathematical Sciences]*. Springer-Verlag, Berlin, 1999. Translated from the 1992 German original and with a note by Norbert Schappacher, With a foreword by G. Harder.

21. R. W. K. Odoni. The Galois theory of iterates and composites of polynomials. *Proc. London Math. Soc. (3)*, 51(3):385–414, 1985.

22. R. W. K. Odoni. On the prime divisors of the sequence $w_{n+1} = 1 + w_1 \cdots w_n$. *J. London Math. Soc. (2)*, 32(1):1–11, 1985.

23. R. W. K. Odoni. Realising wreath products of cyclic groups as Galois groups. *Mathematika*, 35(1):101–113, 1988.

24. J. F. Ritt. Prime and composite polynomials. *Trans. Amer. Math. Soc.*, 23(1):51–66, 1922.

25. J. F. Ritt. Permutable rational functions. *Trans. Amer. Math. Soc.*, 25(3):399–448, 1923.

26. Joseph H. Silverman. *The arithmetic of dynamical systems*, volume 241 of *Graduate Texts in Mathematics*. Springer-Verlag, 2007. To appear.

27. Paul Vojta. *Diophantine approximations and value distribution theory*, volume 1239 of *Lecture Notes in Mathematics*. Springer-Verlag, Berlin, 1987.

Dessins D'enfants for Single-Cycle Belyi Maps

Michelle Manes, Gabrielle Melamed, and Bella Tobin

Abstract Riemann's Existence Theorem gives the following bijections:

(1) Isomorphism classes of *Belyi maps* of degree d.
(2) Equivalence classes of *generating systems* of degree d.
(3) Isomorphism classes of *dessins d'enfants* with d edges.

In previous work, the first author and collaborators exploited the correspondence between Belyi maps and their generating systems to provide explicit equations for two infinite families of dynamical Belyi maps. We complete this picture by describing the dessins d'enfants for these two families.

Keywords Belyi maps · Dessins d'enfants

1 Introduction

Let X be a smooth projective curve. A Belyi map $f : X \to \mathbb{P}^1$ is a finite cover that is ramified only over the points 0, 1, and ∞. The genus of the Belyi map is the genus of the covering curve X. There are multiple ways to realize a Belyi map of degree d (see, for example, [2, 3, 5]):

MM partially supported by the Simons Foundation grant #359721.

M. Manes (✉) · B. Tobin
Department of Mathematics, University of Hawai'i at Mānoa, 2565 McCarthy Mall Keller 401A, Honolulu, HI 96822, USA
e-mail: mmanes@math.hawaii.edu; tobin@math.hawaii.edu; bellatobin@gmail.com

G. Melamed
Department of Mathematics, University of Connecticut, 341 Mansfield Road U1009, Storrs, CT 06269, USA
e-mail: Gabrielle.Melamed@uconn.edu

© The Author(s) and The Association for Women in Mathematics 2019
J. S. Balakrishnan et al. (eds.), *Research Directions in Number Theory*, Association for Women in Mathematics Series 19, https://doi.org/10.1007/978-3-030-19478-9_7

(1) explicitly, as a degree d function between projective curves;
(2) combinatorially, as a generating system of degree d; and
(3) topologically, as a dessin d'enfant with d edges.

A *generating system* of degree d is a triple of permutations $(\sigma_0, \sigma_1, \sigma_\infty) \in S_d^3$ with the property that $\sigma_0 \sigma_1 \sigma_\infty = 1$ and the subgroup $\langle \sigma_0, \sigma_1 \rangle \subseteq S_d$ is transitive.

A *dessin d'enfant* (henceforth *dessin*) is a connected bipartite graph embedded in an orientable surface. The dessin has a fixed cyclic ordering of its edges at each vertex; this manifests as a labeling.

In general, it is a simple matter to describe a dessin from either a generating system or a function. Given $f : X \to \mathbb{P}^1$:

- Define a black vertex for each inverse image of 0.
- Define a white vertex for each inverse image of 1.
- Define edges by the inverse images of the line segment $(0, 1) \in \mathbb{P}^1$.

This process yields a connected bipartite graph. The labeling of the edges arises from the local monodromy around the vertices.

Similarly, given a generating system $(\sigma_0, \sigma_1, \sigma_\infty) \in S_d^3$, we create a dessin with edges labeled $\{1, 2, \ldots, d\}$ via the following recipe:

- Draw a black vertex for each cycle in σ_0 (including the one-cycles). The cycles in σ_0 then give an ordering of edges around each of these vertices.
- Draw a white vertex for each cycle in σ_1 (including the one-cycles). The cycles in σ_1 then give an ordering of edges around each of these vertices.

This determines a bipartite graph. Since σ_0 and σ_1 generate a transitive subgroup of S_d, the graph is connected.

It is equally straightforward to describe a generating system from a dessin. The difficulty in completing the picture is often in giving an explicit function realizing the Belyi map as a covering of $\mathbb{P}^1$. For some recent results in this area, see [4, 6].

In some simple cases, however, we can explicitly realize this triple correspondence for an infinite family of Belyi maps. For example, for pure power maps we have the following:

Belyi map:

$$f : \mathbb{P}^1 \to \mathbb{P}^1$$
$$z \mapsto z^d$$

Generating system:

$$\sigma_0 = d\text{-cycle}, \qquad \sigma_1 = \text{trivial}, \qquad \sigma_\infty = d\text{-cycle}.$$

Dessin d'enfant: See Figure 1

And for Chebyshev polynomials we have

Fig. 1 The dessin for the degree d power map.

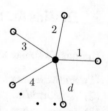

Fig. 2 The Chebyshev dessins: odd d (top) and even d (bottom).

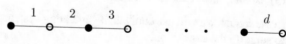

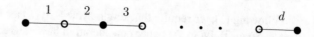

Belyi map:

$$f : \mathbb{P}^1 \to \mathbb{P}^1$$
$$z \mapsto T_d(z)$$

Generating system:

$$\sigma_0 = (23)(45) \cdots ((d-1)d) \qquad \text{or } (23)(45) \cdots ((d-2)(d-1)).$$
$$\sigma_1 = (12)(34) \cdots ((d-2)(d-1)) \qquad \text{or } (12)(34) \cdots ((d-1)d).$$
$$\sigma_\infty = d\text{-cycle}.$$

Dessin d'enfant: See Figure 2.

Motivated by work in arithmetic dynamics, the authors of [1] study *normalized single-cycle dynamical Belyi maps*. The authors begin with a generating system, and they are able to give explicit formulas for two new infinite families of Belyi maps.

In this note, we give a simple description of the dessins for genus 0 single-cycle dynamical Belyi maps. As an application, we describe the dessins for the two infinite families of maps in [1], completing the triptych in these cases.

Acknowledgments. This material is based on work supported by and while the first author served at the National Science Foundation. Any opinion, findings, and conclusions or recommendations expressed in this material are those of the authors and do not necessarily reflect the views of the National Science Foundation.

2 Dessins for Single-Cycle Belyi Maps

Riemann's Existence Theorem gives the following bijections:

(1) *equivalence classes* of generating systems of degree d.
(2) *isomorphism classes* of Belyi maps $f : X \to \mathbb{P}^1$ of degree d.
(3) *isomorphism classes* of dessins d'enfants with d edges.

In the sequel, we will use equivalent generating systems and isomorphic Belyi maps to simplify the exposition.

Definition. *Let $f : X \to \mathbb{P}^1$ be a Belyi map. If f has a single ramification point above each branch point, we say that f is a* **single-cycle Belyi map**.

Following [1], we require that single-cycle Belyi maps have exactly three critical points. So all Belyi maps under consideration have degree $d \geq 3$. If $f : \mathbb{P}^1 \to \mathbb{P}^1$ has only two critical points, then it is isomorphic to a pure power map. We have already described the dessins for this case in the Introduction.

In the genus 0 single-cycle case, the permutations σ_0, σ_1, and σ_∞ each correspond to a single nontrivial cycle. Hence the generating system can be written as (e_0, e_1, e_∞) where e_i gives the length of the nontrivial cycle in σ_i. Equivalently, e_i is the ramification index of the unique critical point above $i \in \mathbb{P}^1$.

The following theorem classifies the dessins d'enfants for all genus 0 single-cycle Belyi maps.

Theorem 1. *Let $f : \mathbb{P}^1 \to \mathbb{P}^1$ be a degree-d single-cycle Belyi map with generating system (e_0, e_1, e_∞). Then f admits a planar dessin d'enfant with:*

- $d - e_1$ *white vertices of degree one connected to a black vertex of degree e_0,*
- $d - e_0$ *black vertices of degree one connected to a white vertex of degree e_1, and*
- $e_0 + e_1 - d$ *edges connecting the black vertex of degree e_0 and the white vertex of degree e_1.*

See Figure 3.

Proof. Let

$$\sigma_0 = (d - e_0 + 1, d - e_0 + 2, \ldots, d) \text{ and } \sigma_1 = (1, 2, \ldots, e_1).$$

It follows from Riemann-Hurwitz that $e_0 + e_1 + e_\infty = 2d + 1$, so

$$d + 1 \leq e_0 + e_1 \leq 2d - 1.$$

Therefore $\langle \sigma_0, \sigma_1 \rangle$ is transitive since $d - e_0 + 1 \leq e_1$.

Fig. 3 The degree d, genus 0 single-cycle dessin with combinatorial type (e_0, e_1, e_∞).

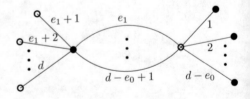

The result then follows immediately from the recipe for producing dessins from generating systems described in the Introduction.   □

Recall that the **diameter** of a graph is the maximal number of vertices traversed in a path. The following result gives a restriction on the diameter of the dessins for all single-cycle Belyi maps (not only the genus 0 case).

Proposition 1. *All single-cycle Belyi maps admit a dessin d'enfant of diameter at most 4.*

Proof. Let $f : X \rightarrow \mathbb{P}^1$ be a single-cycle Belyi map. So there is a unique ramification point above 0 and a unique ramification point above 1. Hence there are exactly two vertices in the dessin with degree greater than 1. This implies that the longest path can only include those two vertices and two additional vertices, one black and one white.   □

3   New Triptychs for Single-Cycle Belyi Maps

Applying Theorem 1 to the Belyi maps in [1] allows us to describe the three-way correspondence for two new infinite families of Belyi maps.

3.1   Single-Cycle Polynomials

Let $f : \mathbb{P}^1 \rightarrow \mathbb{P}^1$ be a degree d single-cycle Belyi polynomial, so $(e_0, e_1, e_\infty) = (d - k, k + 1, d)$ for some $1 \leq k < d - 1$. We have the following correspondence:

Belyi map: (from [1])

$$f : \mathbb{P}^1 \rightarrow \mathbb{P}^1$$

$$z \mapsto cx^{d-k}(a_0x^k + \ldots + a_{k-1}x + a_k),$$

where

$$a_i := \frac{(-1)^{k-i}}{(d-i)} \binom{k}{i} \quad \text{and} \quad c = \frac{1}{k!} \prod_{j=0}^{k}(d - j).$$

Generating system:

$$\sigma_0 = (d - k)\text{-cycle}, \qquad \sigma_1 = (k + 1)\text{-cycle}, \qquad \sigma_\infty = d\text{-cycle}.$$

Dessin d'enfant: See Figure 4.

Example. *The dessin for the polynomial $f(z) = z^3(6z^2 - 15z + 10)$, which has combinatorial type $(3, 3, 5)$, is shown in Figure 5.*

Fig. 4 The dessin with combinatorial type $(d - k, k + 1, d)$.

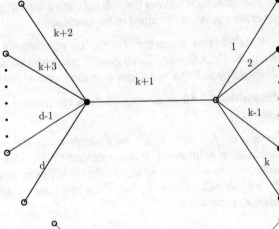

Fig. 5 The dessin with combinatorial type $(3, 3, 5)$.

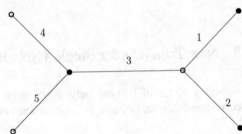

3.2 Symmetric Single-Cycle Belyi Maps

We turn now to the second family of Belyi maps described in [1]. These have combinatorial type $(d - k, 2k + 1, d - k)$, meaning that the critical points above 0 and ∞ have the same ramification index. (Here d is the degree of the Belyi map and $1 \leq k < d - 1$.) We have the following correspondence:

Belyi map: (from [1])

$$f : \mathbb{P}^1 \to \mathbb{P}^1$$

$$z \mapsto x^{d-k} \left(\frac{a_0 x^k - a_1 x^{k-1} + \ldots + (-1)^k a_k}{(-1)^k a_k x^k + \ldots - a_1 x + a_0} \right),$$

where

$$a_i := \binom{k}{i} \prod_{k+i+1 \leq j \leq 2k} (d - j) \prod_{0 \leq j \leq i-1} (d - j) = k! \binom{d}{i} \binom{d - k - i - 1}{k - i}.$$

Generating system:

$$\sigma_0 = (d - k)\text{-cycle}, \qquad \sigma_1 = (2k + 1)\text{-cycle}, \qquad \sigma_\infty = (d - k)\text{-cycle}.$$

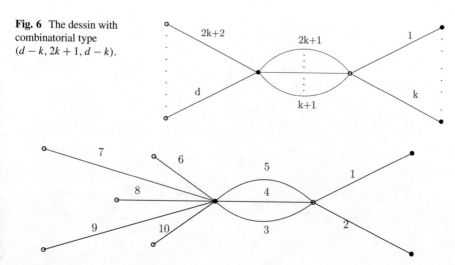

Fig. 6 The dessin with combinatorial type $(d - k, 2k + 1, d - k)$.

Fig. 7 The dessin with combinatorial type $(8, 5, 8)$.

Dessin d'enfant: See Figure 6

Example. *The dessin for the map*

$$f(z) = z^8 \left(\frac{42z^2 - 120z + 90}{90z^2 - 120z + 42} \right),$$

which has combinatorial type $(8, 5, 8)$, *is shown in Figure 7.*

References

1. Jacqueline Anderson, Irene Bouw, Ozlem Ejder, Neslihan Girgin, Valentijn Karemaker, and Michelle Manes. Dynamical Belyi maps, 2017.
2. Frits Beukers and Hans Montanus. Explicit calculation of elliptic fibrations of $K3$-surfaces and their Belyi-maps. In *Number theory and polynomials*, volume 352 of *London Math. Soc. Lecture Note Ser.*, pages 33–51. Cambridge Univ. Press, Cambridge, 2008.
3. Ernesto Girondo and Gabino González-Diez. *Introduction to compact Riemann surfaces and dessins d'enfants*, volume 79 of *London Mathematical Society Student Texts*. Cambridge University Press, Cambridge, 2012.
4. Jeroen Sijsling and John Voight. On computing Belyi maps. In *Numéro consacré au trimestre "Méthodes arithmétiques et applications", automne 2013*, volume 2014/1 of *Publ. Math. Besançon Algèbre Théorie Nr.*, pages 73–131. Presses Univ. Franche-Comté, Besançon, 2014.
5. Leonardo Zapponi. What is...a dessin d'enfant? *Notices of the Amer. Math. Soc.*, 50:788–789, 2003.
6. Alexander Zvonkin. Belyi functions: examples, properties, and applications. http://www.labri.fr/perso/zvonkin/Research/belyi.pdf.

Multiplicative Order and Frobenius Symbol for the Reductions of Number Fields

Antonella Perucca

Abstract Let L/K be a finite Galois extension of number fields, and let G be a finitely generated subgroup of $K^\times$. We study the natural density of the set of primes of K having some prescribed Frobenius symbol in $\mathrm{Gal}(L/K)$, and for which the reduction of G has multiplicative order with some prescribed ℓ-adic valuation for finitely many prime numbers ℓ. This extends in several directions results by Moree and Sury (2009) and by Chinen and Tamura (2012), and has to be compared with the very general result of Ziegler (2006).

Keywords Number fields · Reductions · Multiplicative order · Density · Frobenius symbol

1 Introduction

Consider a Lucas sequence $a^k + b^k$ where $a, b \in \mathbb{Z}$. A prime divisor for the sequence is a prime number that divides at least one term in the sequence. Apart from some trivial cases, counting the prime divisors for the above sequence means counting the prime numbers p such that the multiplicative order of a/b modulo p is even (we may count instead the reductions for which the order is odd). The set of prime divisors admits a natural density, which has been computed by Hasse [3, 4]. There are many related questions, see, for example, the survey by Moree [5].

The following refined question has also been considered: if $L/\mathbb{Q}$ is a finite Galois extension and $\mathfrak{c}$ is a conjugacy class in $\mathrm{Gal}(L/\mathbb{Q})$, how many prime numbers p (unramified in L) are prime divisors for a given Lucas sequence and also fulfill the condition $\mathrm{Frob}_{L/\mathbb{Q}}(p) \subseteq \mathfrak{c}$ for their Frobenius symbol? If L is either quadratic or cyclotomic, the corresponding density has been worked out by Chinen and Tamura [1] and by Moree and Sury [6].

A. Perucca (✉)
University of Luxembourg, Mathematics Research Unit, 6, avenue de la Fonte L-4364, Esch-sur-Alzette, Luxembourg
e-mail: antonella.perucca@uni.lu

© The Author(s) and The Association for Women in Mathematics 2019
J. S. Balakrishnan et al. (eds.), *Research Directions in Number Theory*, Association for Women in Mathematics Series 19, https://doi.org/10.1007/978-3-030-19478-9_8

161

We work in much greater generality. Indeed, we let L/K be any finite Galois extension of number fields and let $\mathfrak{c}$ be a conjugacy-stable subset of the Galois group $\mathrm{Gal}(L/K)$. We also work with any finitely generated subgroup G of $K^\times$. We test coprimality of the order with respect to finitely many prime numbers ℓ and in fact we can also arbitrarily prescribe the ℓ-adic valuation of the order.

Up to excluding finitely many primes $\mathfrak{p}$ of K, we may assume that $\mathfrak{p}$ is unramified in L and that the reduction of G modulo $\mathfrak{p}$ is well-defined.

For simplicity, we now fix one prime number ℓ and count the primes $\mathfrak{p}$ of K that satisfy the following two conditions: firstly, for the Frobenius symbol we must have $\mathrm{Frob}_{L/K}(\mathfrak{p}) \subseteq \mathfrak{c}$; secondly, the order of the group $(G \bmod \mathfrak{p})$ must be coprime to ℓ. Since G only affects the second condition, we may assume w.l.o.g. that G is torsion-free and non-trivial. The general result involves cyclotomic and Kummer extensions (we refer to Section 2 for the definitions and the notation) and it is as follows.

Theorem 1. *Let K be a number field, and fix some non-trivial, finitely generated and torsion-free subgroup G of $K^\times$. Let L be a finite Galois extension of K, and fix some conjugacy-stable subset $\mathfrak{c}$ of $\mathrm{Gal}(L/K)$. Let ℓ be a prime number.*

The set of primes $\mathfrak{p}$ of K satisfying $\mathrm{Frob}_{L/K}(\mathfrak{p}) \subseteq \mathfrak{c}$ and $\ell \nmid \mathrm{ord}(G \bmod \mathfrak{p})$ admits a natural density, which is given by

$$\mathrm{dens}_K(G, \mathfrak{c}, \ell) = \sum_{n=0}^{\infty} \left(\frac{c(n, n)}{[L(n, n) : K]} - \frac{c(n + 1, n)}{[L(n + 1, n) : K]} \right), \qquad (1)$$

where we set $K(m, n) := K(\ell^{-m}1, \ell^{-n}G)$ and $L(m, n) := L \cdot K(m, n)$, and where $c(m, n)$ is the number of elements in $\mathfrak{c}$ that are the identity on $L \cap K(m, n)$.

In the special case $\mathfrak{c} = \mathrm{Gal}(L/K)$, the condition on the Frobenius symbol is trivial and we become the density considered by Debry and Perucca in [2], namely

$$\mathrm{dens}_K(G, \ell) = \sum_{n=0}^{\infty} \left(\frac{1}{[K(n, n) : K]} - \frac{1}{[K(n + 1, n) : K]} \right). \qquad (2)$$

We also extend Theorem 1 by requiring coprimality of the order with respect to finitely many prime numbers, i.e., the order of the reduction of G must be coprime to some given integer $m > 1$. We call the corresponding density $\mathrm{dens}_K(G, \mathfrak{c}, m)$. The density $\mathrm{dens}_K(G, m)$ without the condition on the Frobenius symbol has been considered in [7]. We prove in general that $\mathrm{dens}_K(G, \mathfrak{c}, m)$ is an explicitly computable rational number, see Theorem 17. Finally, rather than considering only the case of valuation 0, for each prime divisor ℓ of m we may arbitrarily prescribe the ℓ-adic valuation of the order of the reduction of G, see Theorem 18.

We conclude by justifying our work in view of the elegant and very general result of Ziegler [8, Theorem 1]. The advantages of our approach are that we work unconditionally, we consider a finitely generated group of any rank, and we show that the density is computable and it is a rational number.

2 Preliminaries

The Frobenius symbol: If L/K is a finite Galois extension of number fields, we define the Frobenius symbol $\mathrm{Frob}_{L/K}(\mathfrak{p})$ of a prime $\mathfrak{p}$ of K that does not ramify in L. Write $\mathcal{O}_K$ and $\mathcal{O}_L$ for the ring of integers of K and L, respectively. The Frobenius symbol of $\mathfrak{p}$ is the conjugacy class in $\mathrm{Gal}(L/K)$ consisting of those σ satisfying the following condition: there exists some prime $\mathfrak{q}$ of L lying over $\mathfrak{p}$ such that for all $\alpha \in \mathcal{O}_L$ we have

$$\sigma(\alpha) \equiv \alpha^{\#(\mathcal{O}_K/\mathfrak{p}\mathcal{O}_K)}(\mathrm{mod}\ \mathfrak{q}).$$

The m-adic valuation (and tuples): We let $m > 1$ be a square-free integer and we consider its prime factorization $m = \ell_1 \cdots \ell_f$. Set $I = \{1, \ldots, f\}$. We define the m-adic valuation as the f-tuple of the ℓ_i-adic valuations where $i \in I$.

If n is a non-negative integer, to ease notation we also write n for the f-tuple with all entries equal to n.

Let A, B be f-tuples. We write $A \leqslant B$ if each entry of A is smaller than or equal to the corresponding entry of B. We write $A < B$ if $A \leqslant B$ and $A \neq B$ hold.

Let A be any f-tuple of non-negative integers, and write A_i for the i-th entry of A. We define

$$m^A := \prod_{i \in I} \ell_i^{A_i}.$$

Torsion and Kummer extensions: Let K be a number field. If n is a positive integer, the n-th cyclotomic extension of K will be denoted by $K(n^{-1}1)$. If G is a finitely generated subgroup of $K^\times$, we consider the Kummer extension $K(n^{-1}G)$, which is the smallest extension of K over which all n-th roots of G (namely all algebraic numbers x such that $x^n \in G$) are defined. When we are interested in powers of some fixed prime number ℓ, we use the following compact notation:

$$K(m, n) := K(\ell^{-m}1, \ell^{-n}G).$$

The union of these fields is usually denoted by $K(\ell^{-\infty}G)$. We will work with some positive square-free number $m > 1$ and then consider $K(m^{-\infty}G)$, i.e., the compositum of the fields $K(\ell^{-\infty}G)$ for ℓ varying among the prime divisors of m.

Lemma 2. *Let K'/K be a finite Galois extension of number fields. Let S (resp. S') be a set of primes of K (resp. K') admitting a Dirichlet density. If S consists only of primes splitting completely in K', and if S' is the set of primes of K' lying above the primes in S, then we have $\mathrm{dens}_K(S) = [K' : K]^{-1} \cdot \mathrm{dens}_{K'}(S').$*

Proof. This fact is well known, see, e.g., [7, Proposition 1] for a proof.   □

3   The Density as an Infinite Sum

Theorem 3. *Let K be a number field, and fix some non-trivial, finitely generated and torsion-free subgroup G of $K^\times$. Let L be a finite Galois extension of K, and fix some conjugacy-stable subset $\mathfrak{c}$ of $\mathrm{Gal}(L/K)$. Let m be the product of f distinct prime numbers. The set of primes $\mathfrak{p}$ of K satisfying $\mathrm{Frob}_{L/K}(\mathfrak{p}) \subseteq \mathfrak{c}$ and such that $\mathrm{ord}(G \bmod \mathfrak{p})$ is coprime to m admits a natural density, which is given by*

$$\mathrm{dens}_K(G, \mathfrak{c}, m) = \sum_A \mathrm{dens}_K(G, \mathfrak{c}, m, A) \tag{3}$$

where A varies over the f-tuples of non-negative integers and where $\mathrm{dens}_K(G, \mathfrak{c}, m, A)$ is the natural density of the set of primes $\mathfrak{p}$ of K that split completely in $K(m^{-A}G)$, do not split completely in $K(m^{-B}1)$ for any $B > A$, and satisfy $\mathrm{Frob}_{L/K}(\mathfrak{p}) \subseteq \mathfrak{c}$.

Proof. Up to excluding finitely many primes of K, we may suppose them to be unramified in L and in $K(m^{-\infty}G)$. The natural density $\mathrm{dens}_K(G, \mathfrak{c}, m, A)$ is well-defined by the Chebotarev Density Theorem. The set of primes considered for $\mathrm{dens}_K(G, \mathfrak{c}, m)$ is the disjoint union over A of those considered for $\mathrm{dens}_K(G, \mathfrak{c}, m, A)$, see [7, Theorem 9]. The natural density $\mathrm{dens}_K(G, \mathfrak{c}, m)$ then exists because the tail of the sum is contained in the set of primes splitting completely in $K(m^{-n}G)$ for some $n \geqslant 1$, and this set has a natural density going to zero for n going to infinity. $\square$

Proof of Theorem 1. By Theorem 3 we only need evaluating $\mathrm{dens}_K(G, \mathfrak{c}, n)$, i.e., consider the set of primes of K that split completely in $K(n, n)$, do not split in $K(n + 1, n)$, and satisfy the condition on the Frobenius symbol. We first count the primes $\mathfrak{p}$ of K whose Frobenius symbol is in $\mathfrak{c}$ and split completely in $K(n, n)$ and then remove those whose Frobenius symbol is in $\mathfrak{c}$ and split completely in $K(n + 1, n)$. Let $N \in \{n, n + 1\}$. By the Chebotarev Density Theorem, we only have to evaluate the relative size of the conjugacy-stable subset of $\mathrm{Gal}(L(N, n)/K)$ consisting of those elements that map to $\mathfrak{c}$ in $\mathrm{Gal}(L/K)$ and map to the identity on $K(N, n)$. The fields L and $K(N, n)$ are linearly disjoint over their intersection and their compositum is $L(N, n)$. Up to considering a factor $[L(N, n) : L]^{-1}$ it is then equivalent to compute the relative size of the conjugacy class in $\mathrm{Gal}(L/K)$ consisting of those elements in $\mathfrak{c}$ that are the identity on $L \cap K(N, n)$. The latter is $c(N, n)/[L : K]$ and we conclude. $\square$

4 General Remarks

We keep the notation of Theorem 3 and investigate $\text{dens}_K(G, \mathfrak{c}, m)$. Recall that we write $\text{dens}_K(G, m)$ for the density analogous to $\text{dens}_K(G, \mathfrak{c}, m)$ where we neglect the condition on the Frobenius symbol.

Lemma 4. *The following assertions hold:*

(a) *We may replace the field L by L' and $\mathfrak{c}$ by $\mathfrak{c}'$, where L'/L is any finite Galois extension and $\mathfrak{c}'$ is the preimage of $\mathfrak{c}$ in $\text{Gal}(L'/K)$ because we have*

$$\text{dens}_K(G, \mathfrak{c}, m) = \text{dens}_K(G, \mathfrak{c}', m).$$

(b) *We may partition $\mathfrak{c}$ into sets which are the union of conjugacy classes and add together the densities calculated with respect to each element in the partition.*

(c) *If two conjugacy-stable subsets $\mathfrak{c}$ and $\mathfrak{c}'$ give a partition of $\text{Gal}(L/K)$, then we may compute $\text{dens}_K(G, \mathfrak{c}, m)$ from $\text{dens}_K(G, \mathfrak{c}', m)$.*

(d) *We may reduce w.l.o.g. to the following special case: for every Galois subextension F/K of L/K either all elements of $\mathfrak{c}$ are the identity on F or none is (which possibility applies depends on F).*

Proof. (a) The Frobenius symbol w.r.t L/K lies in $\mathfrak{c}$ if and only if the Frobenius symbol w.r.t L'/K lies in $\mathfrak{c}'$. (b) Obvious. (c) We have

$$\text{dens}_K(G, \mathfrak{c}, m) = \text{dens}_K(G, m) - \text{dens}_K(G, \mathfrak{c}', m)$$

where $\text{dens}_K(G, m)$ is the density without considering the Frobenius condition, which can be explicitly computed by the results in [7]. (d) The kernel of the quotient map $\text{Gal}(L/K) \to \text{Gal}(F/K)$ and its complement induce a partition on $\mathfrak{c}$, and we may apply (b). Since L/K has only finitely many Galois subextensions, we may repeat this procedure and get the assertion for every F. $\qquad\square$

Remark 5. *We may always reduce to the case $L \subset K(m^{-\infty}G)$ by combining Lemma 4(b) with the following result.*

Proposition 6. *Let $L' = L \cap K(m^{-\infty}G)$ and let $\mathfrak{c}'$ be the projection of $\mathfrak{c}$ in $\text{Gal}(L'/K)$.*

(a) *If each element in $\mathfrak{c}'$ has the same amount of preimages in $\mathfrak{c}$, then we have*

$$\text{dens}_K(G, \mathfrak{c}, m) = \frac{\#\mathfrak{c}}{\#\mathfrak{c}' \cdot [L : L']} \cdot \text{dens}_K(G, \mathfrak{c}', m). \tag{4}$$

In particular, if $\mathfrak{c}$ is the inverse image of $\mathfrak{c}'$ in $\text{Gal}(L/K)$, then we have

$$\text{dens}_K(G, \mathfrak{c}, m) = \text{dens}_K(G, \mathfrak{c}', m). \tag{5}$$

(b) *We may partition* $\mathfrak{c}$ *into conjugacy classes for which part (a) applies, and whose images in* $\mathrm{Gal}(L'/K)$ *give a partition of* $\mathfrak{c}'$.

Proof. **(a)** Call $\mathfrak{c}'' \supseteq \mathfrak{c}$ the inverse image of $\mathfrak{c}'$ in $\mathrm{Gal}(L/K)$. By Lemma 4(a) we have to prove

$$\mathrm{dens}_K(G, \mathfrak{c}, m) = \frac{\#\mathfrak{c}}{\#\mathfrak{c}''} \cdot \mathrm{dens}_K(G, \mathfrak{c}'', m).$$

By Theorem 3 it then suffices to fix any f-tuple A of non-negative integers and show that

$$\mathrm{dens}_K(G, \mathfrak{c}, m, A) = \frac{\#\mathfrak{c}}{\#\mathfrak{c}''} \cdot \mathrm{dens}_K(G, \mathfrak{c}'', m, A)$$

holds. Whether a prime of K has to be counted or not for these two densities only depends on its Frobenius class in the extension $L(m^{-A+1}G)/K$, therefore we may apply the Chebotarev Density Theorem to this finite Galois extension. By the assumption on $\mathfrak{c}'$, the sizes of the conjugacy-stable sets corresponding to the two densities have ratio $\#\mathfrak{c}/\#\mathfrak{c}''$, and we conclude. **(b)** It suffices to remark that two elements of $\mathrm{Gal}(L/K)$ having the same restriction to L' are mapped under conjugation to two elements with the same property. $\square$

The following corollary deals with the generic case, in which the extensions L and $K(m^{-\infty}G)$ are linearly disjoint. The condition on the Frobenius symbol provides the term $\frac{\#\mathfrak{c}}{[L:K]}$ and the condition on the order gives the remaining term (the two conditions are here independent).

Corollary 7. *If* $L \cap K(m^{-\infty}G) = K$, *then we have*

$$\mathrm{dens}_K(G, \mathfrak{c}, m) = \frac{\#\mathfrak{c}}{[L : K]} \cdot \mathrm{dens}_K(G, m). \tag{6}$$

Proof. This is Proposition 6(a) where we set $L' = K$. $\square$

Lemma 8. *We may reduce w.l.o.g. to the special case* $K = K(m^{-1}1)$ *(because if* ℓ *is a prime divisor of m we may either recover* $\mathrm{dens}_K(G, \mathfrak{c}, m)$ *from the analogue density computed over* $K(\ell^{-1}1)$ *or we may replace m by* $\frac{m}{\ell}$). *In particular, if* ℓ *varies over the prime divisors of m, we may reduce w.l.o.g. to the case where the extensions* $K(\ell^{-\infty}G)$ *are linearly disjoint over K.*

Proof. By Lemma 4(a) we may suppose that L contains $K(m^{-1}1)$. By Lemma 4(d), for each prime divisor ℓ of m we may suppose that either $\mathfrak{c} \subseteq \mathrm{Gal}(L/K(\ell^{-1}1))$ or that $\mathfrak{c} \cap \mathrm{Gal}(L/K(\ell^{-1}1)) = \emptyset$. In the first case we have

$$\mathrm{dens}_K(G, \mathfrak{c}, m) = [K(\ell^{-1}1) : K]^{-1} \cdot \mathrm{dens}_{K(\ell^{-1}1)}(G, \mathfrak{c}, m)$$

by Lemma 2. In the second case the coprimality condition w.r.t. ℓ is trivial (being a consequence of the condition on the Frobenius symbol) so we may replace m by $\frac{m}{\ell}$. For the second assertion notice that the Galois group of $K(\ell^{-\infty}G)$ over K is a pro-ℓ-group if $K = K(m^{-1}1)$.   □

5 Examples and Special Cases

Proposition 9 (Intermediate Galois Groups; the Identity). *Suppose that* $\mathfrak{c} = \mathrm{Gal}(L/K')$ *holds for some intermediate extension* $K \subseteq K' \subseteq L$ *which is Galois over* K. *Then we have*

$$\mathrm{dens}_K(G, \mathfrak{c}, m) = \frac{1}{[K' : K]} \cdot \mathrm{dens}_{K'}(G, m), \tag{7}$$

In particular, we have

$$\mathrm{dens}_K(G, \{\mathrm{id}_L\}, m) = \frac{1}{[L : K]} \cdot \mathrm{dens}_L(G, m). \tag{8}$$

Proof. We are interested in the primes $\mathfrak{p}$ of K that split completely in K' and for which the order of $(G \bmod \mathfrak{p})$ is coprime to m. This amounts to counting the primes $\mathfrak{q}$ of K' such that the order of $(G \bmod \mathfrak{q})$ is coprime to m. Finally, we can apply Lemma 2. The second assertion is the special case $K' = L$.   □

Remark 10 (Abelian Extensions). *If the extension* L/K *is abelian, then by Lemma 4(b) we may suppose that* $\mathfrak{c}$ *consists of one element. More generally, we may consider* $\mathfrak{c} = \{\sigma\}$ *for any* σ *in the center of* $\mathrm{Gal}(L/K)$. *By Proposition 9 we may suppose that* σ *is not the identity.*

Remark 11 (Quadratic Extensions). *If* L/K *is a quadratic extension, there are three possibilities for* $\mathfrak{c}$:

(1) If $\mathfrak{c} = \mathrm{Gal}(L/K)$, *then we have* $\mathrm{dens}_K(G, \mathfrak{c}, m) = \mathrm{dens}_K(G, m)$.

(2) If $\mathfrak{c} = \{\mathrm{id}_L\}$, *Proposition 9 gives* $\mathrm{dens}_K(G, \mathfrak{c}, m) = \frac{1}{2} \mathrm{dens}_L(G, m)$.

(3) If $\mathfrak{c} = \mathrm{Gal}(L/K) \setminus \{\mathrm{id}_L\}$, *then Lemma 4(c) and the previous assertions give*

$$\mathrm{dens}_K(G, \mathfrak{c}, m) = \mathrm{dens}_K(G, m) - \frac{1}{2} \mathrm{dens}_L(G, m).$$

Remark 12 (Cyclotomic Extensions). *If* L/K *is a cyclotomic extension, i.e., if* $L = K(n^{-1}1)$ *for some* $n \geqslant 1$, *then in particular we have an abelian Galois extension. We may then suppose by Lemma 4(b) and Proposition 9 that* $\mathfrak{c} = \{\sigma\}$ *where* σ *is not the identity: if* ζ_n *is a primitive n-th root of unity, we have* $\sigma(\zeta_n) = \zeta_n^s$ *for some integer* $1 \leqslant s < n$ *coprime to n. Thus the condition on* *the Frobenius symbol means considering the primes of K lying above the prime numbers congruent to s modulo n.*

Proposition 13 (Trivial Coprimality Condition). *Let ℓ vary over the prime divisors of m. If, for every ℓ, no element of c is the identity on $L \cap K(\ell^{-1}1)$, then we have*

$$\mathrm{dens}_K(G, c, m) = \frac{\#c}{[L : K]}. \tag{9}$$

Proof. The primes $\mathfrak{p}$ of K satisfying $\mathrm{Frob}_{L/K}(\mathfrak{p}) \subseteq c$ are such that ℓ does not divide the order of the multiplicative group of the residue field at $\mathfrak{p}$. In particular, the order of $(G \bmod \mathfrak{p})$ is coprime to ℓ. The assertion is then a consequence of the Chebotarev Density Theorem. □

We now investigate the formula of Theorem 1 by choosing the field L and the conjugacy-stable set c in different ways with respect to the involved cyclotomic/Kummer extensions. Remark that the summands of (1) are non-negative.

Example 14. Let ℓ be a prime number. Taking $L = K(\ell^{-t}1)$ and $c = \{\mathrm{id}_L\}$ for some integer $t \geqslant 1$ gives a density $D(t) := \mathrm{dens}_K(G, c, \ell)$ which is non-increasing with t, and that eventually is decreasing with t. We can write

$$D(t) = \sum_{n=t}^{\infty} \left(\frac{1}{[L(n, n) : K]} - \frac{1}{[L(n+1, n) : K]} \right). \tag{10}$$

Set $\ell = 3$. Let us fix K such that $K \cap \mathbb{Q}(3^{-\infty}1) = \mathbb{Q}$, and choose G of positive rank r such that $K(3^{-\infty}G)$ has maximal degree over $K(3^{-\infty}1)$. Notice that this choice of K and G corresponds to the generic case. We then have

$$D(t) = \sum_{n=t}^{\infty} \frac{1}{2 \cdot 3^{(n-1)+nr}} - \frac{1}{2 \cdot 3^{n+nr}} = \sum_{n=t}^{\infty} 3^{-n(r+1)} = \frac{1}{3^{(t-1)(r+1)}(3^{r+1} - 1)}.$$

Clearly we could have done similar calculations for a different choice of ℓ.

Example 15. Let ℓ be a prime number, and consider $L = K(\ell^{-t}1)$ for some $t \geqslant 1$. Since the Galois extension L/K is abelian (and because of the previous example) we may fix without loss of generality some integer $0 \leqslant s \leqslant t - 1$ and consider $c = \{\sigma\}$, where σ fixes all ℓ^s-th roots of unity and does not fix the primitive ℓ^{s+1}-th roots of unity. We write $D(t, s) := \mathrm{dens}_K(G, c, \ell)$. The only contribution to the density in (1) is the summand $n = s$. In the generic case, for G of rank r we obtain

$$D(t, s) = \frac{1}{[L(s, s) : K]} = \frac{1}{[K(\ell^{-t}1) : K] \cdot \ell^{rs}}. \tag{11}$$

For $\ell = 3$, $K = \mathbb{Q}$ and $s = t - 1$ the formula gives the density $\frac{1}{2} \cdot 3^{-(r+1)s}$.

6 On the Computability of the Density

We keep the notation of Theorem 3 and investigate the computability and the rationality of the density $\mathrm{dens}_K(G, \mathfrak{c}, m)$. We write $\mathrm{dens}_K(G, m)$ if we neglect the condition on the Frobenius symbol.

- By Remark 5 we may suppose $L \subseteq K(m^{-\infty}G)$.
- We assume w.l.o.g. the uniformity on the subfields provided by Lemma 4(d).
- By Lemma 8 we may suppose that $K = K(m^{-1}1)$ and hence that for ℓ varying in the prime divisors of m the extensions $K(\ell^{-\infty}G)$ are linearly disjoint over K.

Definition (Free Primes, Bounded Primes). Let us fix a conjugacy-stable subset of the Galois group $\mathrm{Gal}(L/K)$, and some square-free integer $m \geqslant 2$. We call a prime divisor ℓ of m *free* if all elements of $\mathfrak{c}$ are the identity on $L \cap K(\ell^{-\infty}G)$, and *bounded* otherwise. We write $m = m_B \cdot m_F$ where m_B is the product of the bounded primes and m_F the product of the free primes. By definition, there is some positive integer n_0 (which we call the *bound*) such that for all bounded primes ℓ all elements of $\mathfrak{c}$ are not the identity on $L \cap K(\ell^{-n_0}G)$.

- By Lemma 2 for each free prime ℓ we may extend the base field to $L \cap K(\ell^{-\infty}G)$, i.e., suppose that $K = L \cap K(\ell^{-\infty}G)$. Then the free primes do not appear in the Frobenius symbol condition.

From Theorem 3 we know

$$\mathrm{dens}_K(G, \mathfrak{c}, m) = \sum_A \mathrm{dens}_K(G, \mathfrak{c}, m, A). \tag{12}$$

We write $A = (X, Y)$ by sorting the indexes for the bounded primes and for the free primes, respectively.

- We may suppose $A \neq Y$ because if $A = Y$, then we have $\mathrm{dens}_K(G, \mathfrak{c}, m) = \mathrm{dens}_K(G, m)$ and the latter density is rational and computable by the results in [2, 7].
- We may suppose $A \neq X$ because if $A = X$, then the density reduces to a finite sum of computable rational numbers. Indeed, we have $\mathrm{dens}_K(G, \mathfrak{c}, m, A) = 0$ if $A \leqslant n_0$ does not hold and hence (12) becomes

$$\mathrm{dens}_K(G, \mathfrak{c}, m) = \sum_{A \leqslant n_0} \mathrm{dens}_K(G, \mathfrak{c}, m, A).$$

- We may then suppose that the decomposition $A = (X, Y)$ is non-trivial.

As an aside remark, notice that by [7, Corollary 12] the density $\mathrm{dens}_K(G, m_F)$ is the product of $\mathrm{dens}_K(G, \ell)$ where ℓ varies over the prime divisors of m_F.

Theorem 16. *With the above restrictions (all of which were made without loss of generality) we can write*

$$\text{dens}_K(G, \mathfrak{c}, m) = \left(\sum_{X \leqslant n_0} \text{dens}_K(G, \mathfrak{c}, m_B, X) \right) \cdot \text{dens}_K(G, m_F). \tag{13}$$

Proof. Consider the notation introduced in this section. If $A = (X, Y)$ and $X \leqslant n_0$ does not hold, then we have $\text{dens}_K(G, \mathfrak{c}, m, A) = 0$. Thus (12) becomes

$$\text{dens}_K(G, \mathfrak{c}, m) = \sum_{X \leqslant n_0} \sum_Y \text{dens}_K(G, \mathfrak{c}, m, (X, Y)). \tag{14}$$

Fix $A = (X, Y)$ and work in $F = L(m^{-A+1}G)$, which is the compositum of the extensions $F_1 = L(m_B^{-(X+1)}G)$ and $F_2 = K(m_F^{-(Y+1)}G)$. By the restrictions that we made, the fields F_1 and F_2 are linearly disjoint over K.

The density $\text{dens}_K(G, \mathfrak{c}, m, (X, Y))$ can be computed by the Chebotarev Density Theorem because it corresponds to some conjugacy-stable subset $\mathfrak{c}_A$ of $\text{Gal}(F/K)$. We can write $\mathfrak{c}_A = \mathfrak{c}_{A,X} \times \mathfrak{c}_{A,Y}$ by identifying $\text{Gal}(F/K)$ and $\text{Gal}(F_1/K) \times \text{Gal}(F_2/K)$.

Considering $\mathfrak{c}_{A,Y}$ in $\text{Gal}(F_2/K)$ gives $\text{dens}_K(G, m_F, Y)$ because by our restrictions the original condition on the Frobenius symbol does not affect the free primes. Considering $\mathfrak{c}_{A,X}$ in $\text{Gal}(F_1/K)$ gives $\text{dens}_K(G, \mathfrak{c}, m_B, X)$. We then deduce

$$\text{dens}_K(G, \mathfrak{c}, m) = \sum_{X \leqslant n_0} \text{dens}_K(G, \mathfrak{c}, m_B, X) \cdot \sum_Y \text{dens}_K(G, m_F, Y),$$

which gives the statement.   □

We also have the following result:

Theorem 17. *The density* $\text{dens}_K(G, \mathfrak{c}, m)$ *is a computable rational number.*

Proof. The finite sum in (13) is computable and gives a rational number because each summand has these properties. The density $\text{dens}_K(G, m_F)$ is an explicitly computable rational number by the results in [2, 7] because there is no Frobenius condition involved in the density.   □

Theorem 18. *Let $m > 1$ be a square-free integer. Consider the set of primes $\mathfrak{p}$ of K such that the multiplicative order of $(G \bmod \mathfrak{p})$ has some prescribed m-adic valuation and such that $\mathfrak{p}$ has some prescribed Frobenius symbol in L/K. The natural density of this set is well-defined, and it is a computable rational number.*

Proof. We have already proven the theorem for the special case where the m-adic valuation is the tuple 0 because that means requiring the order of $(G \bmod \mathfrak{p})$ to be coprime to m. If V is some m-adic valuation, then we can apply the statement for the above known case to $m^V G$ and obtain that the multiplicative order of $(G \bmod \mathfrak{p})$ has m-adic valuation at most V. We may then conclude with the help of the inclusion-exclusion principle.   □

Acknowledgements: The author sincerely thanks Pieter Moree for inviting her to visit the MPIM Bonn in April 2017 and for suggesting the problem which has been solved in this paper.

References

1. K. Chinen and C. Tamura, On a distribution property of the residual order of $a(\text{mod } p)$ with a quadratic residue condition, *Tokyo J. Math.* **35** (2012), 441–459.
2. C. Debry and A. Perucca, *Reductions of algebraic integers*, J. Number Theory, **167** (2016), 259–283.
3. H. Hasse, *Über die Dichte der Primzahlen p, für die eine vorgegebene ganzrationale Zahl $a \neq 0$ von durch eine vorgegebene Primzahl $l \neq 2$ teilbarer bzw. unteilbarer Ordnung mod p ist*, Math. Ann. **162** (1965/1966), 74–76.
4. H. Hasse, *Über die Dichte der Primzahlen p, für die eine vorgegebene ganzrationale Zahl $a \neq 0$ von gerader bzw. ungerader Ordnung mod p ist*, Math. Ann. **166** (1966), 19–23.
5. P. Moree, Artin's primitive root conjecture – a survey, *Integers* **12A** (2012), No. 6, 1305–1416.
6. P. Moree and B. Sury, Primes in a prescribed arithmetic progression dividing the sequence $\{a^k + b^k\}_{k=1}^{\infty}$, *Int. J. Number Theory* **5** (2009), 641–665.
7. A. Perucca, *Reductions of algebraic integers II*, I. I. Bouw et al. (eds.), Women in Numbers Europe II, Association for Women in Mathematics Series 11 (2018), 10–33.
8. V. Ziegler, *On the distribution of the order of number field elements modulo prime ideals.* Unif. Distrib. Theory 1 (2006), no. 1, 65–85.

Quantum Modular Forms and Singular Combinatorial Series with Distinct Roots of Unity

Amanda Folsom, Min-Joo Jang, Sam Kimport, and Holly Swisher

Abstract Understanding the relationship between mock modular forms and quantum modular forms is a problem of current interest. Both mock and quantum modular forms exhibit modular-like transformation properties under suitable subgroups of $SL_2(\mathbb{Z})$, up to nontrivial error terms; however, their domains (the upper half-plane $\mathbb{H}$ and the rationals $\mathbb{Q}$, respectively) are notably different. Quantum modular forms, originally defined by Zagier in 2010, have also been shown to be related to the diverse areas of colored Jones polynomials, meromorphic Jacobi forms, partial theta functions, vertex algebras, and more.

In this paper we study the $(n+1)$-variable combinatorial rank generating function $R_n(x_1, x_2, \ldots, x_n; q)$ for n-marked Durfee symbols. These are $n + 1$ dimensional multi-sums for $n > 1$, and specialize to the ordinary two variable partition rank

Acknowledgements: The authors thank the Banff International Research Station (BIRS) and the Women in Numbers 4 (WIN4) workshop for the opportunity to initiate this collaboration. The first author is grateful for the support of National Science Foundation grant DMS-1449679, and the Simons Foundation.

A. Folsom
Department of Mathematics and Statistics, Amherst College, Amherst, MA 01002, USA
e-mail: afolsom@amherst.edu

M.-J. Jang
Department of Mathematics, The University of Hong Kong, Room 318, Run Run Shaw Building, Pokfulam, Hong Kong
e-mail: min-joo.jang@hku.hk

S. Kimport
Department of Mathematics, Stanford University, 450 Serra Mall, Building 380, Stanford, CA 94305, USA
e-mail: skimport@stanford.edu

H. Swisher (✉)
Department of Mathematics, Oregon State University, Kidder Hall 368, Corvallis, OR 97331, USA
e-mail: swisherh@math.oregonstate.edu

generating function when $n = 1$. The mock modular properties of R_n when viewed as a function of $\tau \in \mathbb{H}$, with $q = e^{2\pi i \tau}$, for various n and fixed parameters $x_1, x_2, \cdots, x_n$, have been studied in a series of papers. Namely, by Bringmann and Ono when $n = 1$ and x_1 a root of unity; by Bringmann when $n = 2$ and $x_1 = x_2 = 1$; by Bringmann, Garvan, and Mahlburg for $n \geq 2$ and $x_1 = x_2 = \cdots = x_n = 1$; and by the first and third authors for $n \geq 2$ and the x_j suitable roots of unity $(1 \leq j \leq n)$.

The quantum modular properties of R_1 readily follow from existing results. Here, we focus our attention on the case $n \geq 2$, and prove for any $n \geq 2$ that the combinatorial generating function R_n is a quantum modular form when viewed as a function of $x \in \mathbb{Q}$, where $q = e^{2\pi i x}$, and the x_j are suitable distinct roots of unity.

Keywords Quantum modular forms · Mock modular forms · Modular forms · Durfee symbols · Combinatorial rank functions · Partitions

1 Introduction and Statement of Results

1.1 Background

A *partition* of a positive integer n is any non-increasing sum of positive integers that adds to n. Integer partitions and modular forms are beautifully and intricately linked, due to the fact that the generating function for the partition function $p(n) :=$ #{partitions of n} is related to Dedekind's eta function $\eta(\tau)$, a weight $\frac{1}{2}$ modular form defined by

$$\eta(\tau) := q^{\frac{1}{24}} \prod_{n=1}^{\infty} (1 - q^n). \tag{1.1}$$

Namely,

$$1 + \sum_{n=1}^{\infty} p(n) q^n = \frac{1}{(q; q)_\infty} = q^{\frac{1}{24}} \eta(\tau)^{-1}, \tag{1.2}$$

where here and throughout this section $q := e^{2\pi i \tau}$, $\tau \in \mathbb{H} := \{x + iy \mid x \in \mathbb{R}, y \in \mathbb{R}^+\}$ the upper half of the complex plane, and the q-Pochhammer symbol is defined for $n \in \mathbb{N}_0 \cup \{\infty\}$ by

$$(a)_n = (a; q)_n := \prod_{j=1}^{n} (1 - aq^{j-1}).$$

In fact, the connections between partitions and modular forms go much deeper, and one example of this is given by the combinatorial rank function. Dyson [10] defined the *rank* of a partition to be its largest part minus its number of parts, and the *partition rank function* is defined by

$$N(m, n) := \#\{\text{partitions of } n \text{ with rank equal to } m\}.$$

For example, $N(7, -2) = 2$, because precisely 2 of the 15 partitions of $n = 7$ have rank equal to -2; these are $2 + 2 + 2 + 1$, and $3 + 1 + 1 + 1 + 1$.

Partition rank functions have a rich history in the areas of combinatorics, q-hypergeometric series, number theory, and modular forms. As one particularly notable example, Dyson conjectured that the rank could be used to combinatorially explain Ramanujan's famous partition congruences modulo 5 and 7; this conjecture was later proved by Atkin and Swinnerton-Dyer [2].

It is well known that the associated two variable generating function for $N(m, n)$ may be expressed as a q-hypergeometric series

$$\sum_{m=-\infty}^{\infty} \sum_{n=0}^{\infty} N(m, n) w^m q^n = \sum_{n=0}^{\infty} \frac{q^{n^2}}{(wq; q)_n (w^{-1}q; q)_n} =: R_1(w; q), \qquad (1.3)$$

noting here that $N(m, 0) = \delta_{m0}$, where δ_{ij} is the Kronecker delta function.

Specializations in the w-variable of the rank generating function have been of particular interest in the area of modular forms. For example, when $w = 1$, we have that

$$R_1(1; q) = 1 + \sum_{n=1}^{\infty} p(n) q^n = q^{\frac{1}{24}} \eta^{-1}(\tau) \qquad (1.4)$$

thus recovering (1.2), which shows that the generating function for $p(n)$ is (essentially[1]) the reciprocal of a weight $\frac{1}{2}$ modular form.

If instead we let $w = -1$, then

$$R_1(-1; q) = \sum_{n=0}^{\infty} \frac{q^{n^2}}{(-q; q)_n^2} =: f(q). \qquad (1.5)$$

The function $f(q)$ is not a modular form, but one of Ramanujan's original third order mock theta functions.

Mock theta functions, and more generally mock modular forms and harmonic Maass forms have been major areas of study. In particular, understanding how

---

[1]Here and throughout, as is standard in this subject for simplicity's sake, we may slightly abuse terminology and refer to a function as a modular form or other modular object when in reality it must first be multiplied by a suitable power of q to transform appropriately.

Ramanujan's mock theta functions fit into the theory of modular forms was a question that persisted from Ramanujan's death in 1920 until the groundbreaking 2002 thesis of Zwegers [20]: we now know that Ramanujan's mock theta functions, a finite list of curious q-hypergeometric functions including $f(q)$, exhibit suitable modular transformation properties after they are *completed* by the addition of certain nonholomorphic functions. In particular, Ramanujan's mock theta functions are examples of *mock modular forms*, the holomorphic parts of *harmonic Maass forms*. Briefly speaking, harmonic Maass forms, originally defined by Bruiner and Funke [8], are nonholomorphic generalizations of ordinary modular forms that in addition to satisfying appropriate modular transformations, must be eigenfunctions of a certain weight k-Laplacian operator, and satisfy suitable growth conditions at cusps (see [4, 8, 16, 18] for more).

Given that specializing R_1 at $w = \pm 1$ yields two different modular objects, namely an ordinary modular form and a mock modular form as seen in (1.4) and (1.5), it is natural to ask about the modular properties of R_1 at other values of w. Bringmann and Ono answered this question in [6], and used the theory of harmonic Maass forms to prove that upon specialization of the parameter w to complex roots of unity not equal to 1, the rank generating function R_1 is also a mock modular form. (See also [18] for related work.)

Theorem ([6] Theorem 1.1). If $0 < a < c$, then

$$q^{-\frac{\ell_c}{24}} R_1(\zeta_c^a; q^{\ell_c}) + \frac{i \sin\left(\frac{\pi a}{c}\right) \ell_c^{\frac{1}{2}}}{\sqrt{3}} \int_{-\overline{\tau}}^{i\infty} \frac{\Theta\left(\frac{a}{c}; \ell_c \rho\right)}{\sqrt{-i(\tau + \rho)}} d\rho$$

is a harmonic Maass form of weight $\frac{1}{2}$ on Γ_c.

Here, $\zeta_c^a := e^{\frac{2\pi i a}{c}}$ is a c-th root of unity, $\Theta\left(\frac{a}{c}; \ell_c \tau\right)$ is a certain weight $3/2$ cusp form, $\ell_c := \mathrm{lcm}(2c^2, 24)$, and Γ_c is a particular subgroup of $\mathrm{SL}_2(\mathbb{Z})$.

In this paper, as well as in prior work of two of the authors [12], we study the related problem of understanding the modular properties of certain combinatorial q-hypergeometric series arising from objects called n-marked Durfee symbols, originally defined by Andrews in his notable work [1].

To understand n-marked Durfee symbols, we first describe Durfee symbols. For each partition, the Durfee symbol catalogs the size of its Durfee square, as well as the length of the columns to the right as well as the length of the rows beneath the Durfee square. For example, below we have the partitions of 4, followed by their Ferrers diagrams with any element belonging to their Durfee squares marked by a square (■), followed by their Durfee symbols.

$$4 \qquad 3+1 \qquad 2+2 \quad 2+1+1 \quad 1+1+1+1$$

$$\begin{pmatrix} 1\ 1\ 1 \\ \end{pmatrix}_1, \quad \begin{pmatrix} 1\ 1 \\ 1 \end{pmatrix}_1, \quad ()_2 \quad \begin{pmatrix} 1 \\ 1\ 1 \end{pmatrix}_1, \quad \begin{pmatrix} \\ 1\ 1\ 1 \end{pmatrix}_1,$$

Andrews defined the *rank* of a Durfee symbol to be the length of the partition in the top row, minus the length of the partition in the bottom row. Notice that this gives Dyson's original rank of the associated partition. Andrews refined this idea by defining n-marked Durfee symbols, which use n copies of the integers. For example, the following is a 3-marked Durfee symbol of 55, where α^j, β^j indicate the partitions in their respective columns.

$$\begin{pmatrix} 4_3\ 4_3 & 3_2\ 3_2\ 2_2 & 2_1 \\ 5_3 & 3_2\ 2_2 & 2_1\ 2_1 \end{pmatrix}_5 =: \begin{pmatrix} \alpha^3 & \alpha^2 & \alpha^1 \\ \beta^3 & \beta^2 & \beta^1 \end{pmatrix}_5,$$

Each n-marked Durfee symbol has n ranks, one defined for each column. Let $\text{len}(\pi)$ denote the length of a partition π. Then the nth rank is defined to be $\text{len}(\alpha^n) - \text{len}(\beta^n)$, and each jth rank for $1 \le j < n$ is defined by $\text{len}(\alpha^j) - \text{len}(\beta^j) - 1$. Thus the above example has 3rd rank equal to 1, 2nd rank equal to 0, and 1st rank equal to -1.

Let $\mathcal{D}_n(m_1, m_2, \ldots, m_n; r)$ denote the number of n-marked Durfee symbols arising from partitions of r with ith rank equal to m_i. In [1], Andrews showed that the $(n + 1)$-variable rank generating function for Durfee symbols may be expressed in terms of certain q-hypergeometric series, analogous to (1.3). To describe this, for $n \ge 2$, define

$$R_n(x; q) := \tag{1.6}$$

$$\sum_{\substack{m_1 > 0 \\ m_2, \ldots, m_n \ge 0}} \frac{q^{(m_1 + m_2 + \cdots + m_n)^2 + (m_1 + \cdots + m_{n-1}) + (m_1 + \cdots + m_{n-2}) + \cdots + m_1}}{(x_1 q; q)_{m_1}\left(\frac{q}{x_1}; q\right)_{m_1}(x_2 q^{m_1}; q)_{m_2+1}\left(\frac{q^{m_1}}{x_2}; q\right) \cdots (x_n q^{m_1 + \ldots + m_{n-1}}; q)_{m_n+1}\left(\frac{q^{m_1 + \cdots + m_{n-1}}}{x_n}; q\right)_{m_n+1}},$$

where $x = x_n := (x_1, x_2, \ldots, x_n)$. For $n = 1$, the function $R_1(x; q)$ is defined as the q-hypergeometric series in (1.3). In what follows, for ease of notation, we may also write $R_1(x; q)$ to denote $R_1(x; q)$, with the understanding that $x := x$. In [1], Andrews established the following result, generalizing (1.3).

Theorem ([1] Theorem 10). For $n \ge 1$ we have that

$$\sum_{m_1, m_2, \ldots, m_n = -\infty}^{\infty} \sum_{r=0}^{\infty} \mathcal{D}_n(m_1, m_2, \ldots, m_n; r) x_1^{m_1} x_2^{m_2} \cdots x_n^{m_n} q^r = R_n(x; q).$$

$$\tag{1.7}$$

When $n = 1$, one recovers Dyson's rank, that is, $\mathcal{D}_1(m_1; r) = N(m_1, r)$, so that (1.7) reduces to (1.3) in this case. The mock modularity of the associated two variable generating function $R_1(x_1; q)$ was established in [6] as described in the Theorem above. When $n = 2$, the modular properties of $R_2(1, 1; q)$ were originally studied by Bringmann in [3], who showed that

$$R_2(1, 1; q) := \frac{1}{(q; q)_\infty} \sum_{m \neq 0} \frac{(-1)^{m-1} q^{3m(m+1)/2}}{(1 - q^m)^2}$$

is a *quasimock theta function*. In [5], Bringmann, Garvan, and Mahlburg showed more generally that $R_n(1, 1, \ldots, 1; q)$ is a quasimock theta function for $n \geq 2$. (See [3, 5] for precise details of these statements.)

In [12], two of the authors established the automorphic properties of $R_n(\boldsymbol{x}; q)$, for more arbitrary parameters $\boldsymbol{x} = (x_1, x_2, \ldots, x_n)$, thereby treating families of n-marked Durfee rank functions with additional singularities beyond those of $R_n(1, 1, \ldots, 1; q)$. We point out that the techniques of Andrews [1] and Bringmann [3] were not directly applicable in this setting due to the presence of such additional singularities. These singular combinatorial families are essentially mixed mock and quasimock modular forms. To precisely state a result from [12] along these lines, we first introduce some notation, which we also use for the remainder of this paper. Namely, we consider functions evaluated at certain length n vectors $\boldsymbol{\zeta}_n$ of roots of unity defined as follows (as in [12]).

In what follows, we let n be a fixed integer satisfying $n \geq 2$. Suppose for $1 \leq j \leq n$, $\alpha_j \in \mathbb{Z}$ and $\beta_j \in \mathbb{N}$, where $\beta_j \nmid \alpha_j$, $\beta_j \nmid 2\alpha_j$, and that $\frac{\alpha_r}{\beta_r} \pm \frac{\alpha_s}{\beta_s} \notin \mathbb{Z}$ if $1 \leq r \neq s \leq n$. Let

$$\boldsymbol{\alpha}_n := \left(\frac{\alpha_1}{\beta_1}, \frac{\alpha_2}{\beta_2}, \ldots, \frac{\alpha_n}{\beta_n}\right) \in \mathbb{Q}^n$$

$$\boldsymbol{\zeta}_n := \left(\zeta_{\beta_1}^{\alpha_1}, \zeta_{\beta_2}^{\alpha_2}, \ldots, \zeta_{\beta_n}^{\alpha_n}\right) \in \mathbb{C}^n. \tag{1.8}$$

Remark 1.1. We point out that the dependence of the vector $\boldsymbol{\zeta}_n$ on n is reflected only in the length of the vector, and not (necessarily) in the roots of unity that comprise its components. In particular, the vector components may be chosen to be m-th roots of unity for different values of m.

Remark 1.2. The conditions stated above for $\boldsymbol{\zeta}_n$, as given in [12], do not require $\gcd(\alpha_j, \beta_j) = 1$. Instead, they merely require that $\frac{\alpha_j}{\beta_j} \notin \frac{1}{2}\mathbb{Z}$. Without loss of generality, we will assume here that $\gcd(\alpha_j, \beta_j) = 1$. Then, requiring that $\beta_j \nmid 2\alpha_j$ is the same as saying $\beta_j \neq 2$.

In [12], the authors proved that (under the hypotheses for $\boldsymbol{\zeta}_n$ given above) the completed nonholomorphic function

$$\widehat{\mathcal{A}}(\boldsymbol{\zeta}_n; q) = q^{-\frac{1}{24}} R_n(\boldsymbol{\zeta}_n; q) + \mathcal{A}^-(\boldsymbol{\zeta}_n; q) \tag{1.9}$$

transforms like a modular form. Here the nonholomorphic part $\mathcal{A}^-$ is defined by

$$\mathcal{A}^-(\zeta_n; q) := \frac{1}{\eta(\tau)} \sum_{j=1}^{n} (\zeta_{2\beta_j}^{-3\alpha_j} - \zeta_{2\beta_j}^{-\alpha_j}) \frac{\mathcal{R}_3^-\left(\frac{\alpha_j}{\beta_j}, -2\tau; \tau\right)}{\Pi_j^\dagger(\alpha_n)}, \qquad (1.10)$$

where $\mathcal{R}_3$ is defined in (2.4), and the constant $\Pi_j^\dagger$ is defined in [12, (4.2), with $n \mapsto j$ and $k \mapsto n$]. Precisely, we have the following special case of a theorem established by two of the authors in [12].

Theorem ([12] Theorem 1.1). If $n \geq 2$ is an integer, then $\widehat{\mathcal{A}}(\zeta_n; q)$ is a nonholomorphic modular form of weight $1/2$ on Γ_n with character χ_γ^{-1}.

Here, the subgroup $\Gamma_n \subseteq \mathrm{SL}_2(\mathbb{Z})$ under which $\widehat{\mathcal{A}}(\zeta_n; q)$ transforms is defined by

$$\Gamma_n := \bigcap_{j=1}^{n} \Gamma_0\left(2\beta_j^2\right) \cap \Gamma_1(2\beta_j), \qquad (1.11)$$

and the Nebentypus character χ_γ is given in Lemma 2.1.

1.2 Quantum Modular Forms

In this paper, we study the quantum modular properties of the $(n + 1)$-variable rank generating function for n-marked Durfee symbols $R_n(x; q)$. Loosely speaking, a quantum modular form is similar to a mock modular form in that it exhibits a modular-like transformation with respect to the action of a suitable subgroup of $\mathrm{SL}_2(\mathbb{Z})$; however, the domain of a quantum modular form is not the upper half-plane $\mathbb{H}$, but rather the set of rationals $\mathbb{Q}$ or an appropriate subset. The formal definition of a quantum modular form was originally introduced by Zagier in [19] and has been slightly modified to allow for half-integral weights, subgroups of $\mathrm{SL}_2(\mathbb{Z})$, etc. (see [4]).

Definition 1.3. A weight $k \in \frac{1}{2}\mathbb{Z}$ quantum modular form is a complex-valued function f on $\mathbb{Q}$, such that for all $\gamma = \left(\begin{smallmatrix} a & b \\ c & d \end{smallmatrix}\right) \in \mathrm{SL}_2(\mathbb{Z})$, the functions $h_\gamma :$ $\mathbb{Q} \setminus \gamma^{-1}(i\infty) \to \mathbb{C}$ defined by

$$h_\gamma(x) := f(x) - \varepsilon^{-1}(\gamma)(cx + d)^{-k} f\left(\frac{ax + b}{cx + d}\right)$$

satisfy a "suitable" property of continuity or analyticity in a subset of $\mathbb{R}$.

Remark 1.4. The complex numbers $\varepsilon(\gamma)$, which satisfy $|\varepsilon(\gamma)| = 1$, are such as those appearing in the theory of half-integral weight modular forms.

Remark 1.5. We may modify Definition 1.3 appropriately to allow transformations on subgroups of $SL_2(\mathbb{Z})$. We may also restrict the domains of the functions h_γ to be suitable subsets of $\mathbb{Q}$.

The subject of quantum modular forms has been widely studied since the time of origin of the above definition. For example, quantum modular forms have been shown to be related to the diverse areas of Maass forms, Eichler integrals, partial theta functions, colored Jones polynomials, meromorphic Jacobi forms, and vertex algebras, among other things (see [4] and the references therein). In particular, the notion of a quantum modular form is now known to have direct connection to Ramanujan's original definition of a mock theta function. Namely, in his last letter to Hardy, Ramanujan examined the asymptotic difference between mock theta and modular theta functions as q tends towards roots of unity ζ radially within the unit disc (equivalently, as τ approaches rational numbers vertically in the upper half plane, with $q = e^{2\pi i \tau}$, $\tau \in \mathbb{H}$), and we now know that these radial limit differences are equal to special values of quantum modular forms at rational numbers (see [4, 7, 13]).

1.3 Results

On one hand, exploring the quantum modular properties of the rank generating function for n-marked Durfee symbols R_n in (1.7) is a natural problem given that two of the authors have established automorphic properties of this function on $\mathbb{H}$ (see [12, Theorem 1.1] above), that $\mathbb{Q}$ is a natural boundary to $\mathbb{H}$, and that there has been much progress made in understanding the relationship between quantum modular forms and mock modular forms recently [4]. Moreover, given that R_n is a vast generalization of the two variable rank generating function in (1.3)—both a combinatorial q-hypergeometric series and a mock modular form—understanding its automorphic properties in general is of interest. On the other hand, there is no reason to a priori expect R_n to converge on $\mathbb{Q}$, let alone exhibit quantum modular properties there. Nevertheless, we establish quantum modular properties for the rank generating function for n-marked Durfee symbols R_n in this paper.

For the remainder of this paper, we use the notation

$$\mathcal{V}_n(\tau) := \mathcal{V}(\zeta_n; q),$$

where $\mathcal{V}$ may refer to any one of the functions $\widehat{A}, A^-, R_n$, etc. Moreover, we will write

$$\mathcal{A}_n(\tau) = q^{-\frac{1}{24}} R_n(\zeta_n; q) \tag{1.12}$$

for the holomorphic part of $\widehat{A}$; from [12, Theorem 1.1] above, we have that this function is a mock modular form of weight $1/2$ with character χ_γ^{-1} (see Lemma 2.1) for the group Γ_n defined in (1.11). Here, we will show that $\mathcal{A}_n$ is also a quantum modular form, under the action of a subgroup $\Gamma_{\zeta_n} \subseteq \Gamma_n$ defined in (1.15), with quantum set

$$Q_{\zeta_n} := \left\{ \frac{h}{k} \in \mathbb{Q} \;\middle|\; \begin{array}{l} h \in \mathbb{Z}, k \in \mathbb{N}, \gcd(h,k)=1, \; \beta_j \nmid k \; \forall \, 1 \le j \le n, \\[4pt] \left| \frac{\alpha_j}{\beta_j} k - \left[\frac{\alpha_j}{\beta_j} k \right] \right| > \frac{1}{6} \; \forall \, 1 \le j \le n \end{array} \right\},$$

where $[x]$ is the closest integer to x.

(1.13)

Remark 1.6. For $x \in \frac{1}{2} + \mathbb{Z}$, different sources define $[x]$ to mean either $x - \frac{1}{2}$ or $x + \frac{1}{2}$. The definition of Q_{ζ_n} involving $[\cdot]$ is well-defined for either of these conventions in the case of $x \in \frac{1}{2} + \mathbb{Z}$, as $|x - [x]| = \frac{1}{2}$.

To define the exact subgroup under which $\mathcal{A}_n$ transforms as a quantum automorphic object, we let

$$\ell = \ell(\zeta_n) := \begin{cases} 6\,[\mathrm{lcm}(\beta_1, \ldots, \beta_n)]^2 & \text{if } 3 \nmid \beta_j \text{ for all } 1 \le j \le n, \\ 2\,[\mathrm{lcm}(\beta_1, \ldots, \beta_n)]^2 & \text{if } 3 \mid \beta_j \text{ for some } 1 \le j \le n, \end{cases}$$

(1.14)

and let $S_\ell := \left(\begin{smallmatrix} 1 & 0 \\ \ell & 1 \end{smallmatrix} \right)$, $T := \left(\begin{smallmatrix} 1 & 1 \\ 0 & 1 \end{smallmatrix} \right)$. We then define the group generated by these two matrices as

$$\Gamma_{\zeta_n} := \langle S_\ell, T \rangle.$$

(1.15)

We now state our first main result, which proves that $\mathcal{A}_n(x)$, and hence $e(-\frac{x}{24}) R_n(\zeta_n; e(x))$ is a quantum modular form on Q_{ζ_n} with respect to Γ_{ζ_n}. Here and throughout we let $e(x) := e^{2\pi i x}$.

Theoerm 1.7. *Let $n \ge 2$. For all $\gamma = \left(\begin{smallmatrix} a & b \\ c & d \end{smallmatrix} \right) \in \Gamma_{\zeta_n}$, and $x \in Q_{\zeta_n}$,*

$$H_{n,\gamma}(x) := \mathcal{A}_n(x) - \chi_\gamma (cx+d)^{-\frac{1}{2}} \mathcal{A}_n(\gamma x)$$

is defined, and extends to an analytic function in x on $\mathbb{R} - \{\frac{-c}{d}\}$. In particular, for the matrix S_ℓ,

$$H_{n,S_\ell}(x) = \frac{\sqrt{3}}{2} \sum_{j=1}^{n} \frac{(\zeta_{2\beta_j}^{\alpha_j} - \zeta_{2\beta_j}^{3\alpha_j})}{\Pi_j^\dagger(\alpha_n)} \left[\sum_{\pm} \zeta_6^{\pm 1} \int_{\frac{1}{\ell}}^{i\infty} \frac{g_{\pm\frac{1}{3}+\frac{1}{2}, -\frac{3\alpha_j}{\beta_j}+\frac{1}{2}}(3\rho)}{\sqrt{-i(\rho+x)}} \, d\rho \right]$$

$$+ \sum_{j=1}^{n} \frac{(\zeta_{2\beta_j}^{-3\alpha_j} - \zeta_{2\beta_j}^{-\alpha_j})}{\Pi_j^\dagger(\alpha_n)} (\ell x + 1)^{-\frac{1}{2}} \zeta_{24}^{-\ell} \mathcal{E}_1\left(\frac{\alpha_j}{\beta_j}, \ell; x \right),$$

where the weight $3/2$ *theta functions* $g_{a,b}$ *are defined in* (2.5), *and* $\mathcal{E}_1$ *is defined in* (4.3).

Remark 1.8. As mentioned above, the constants $\Pi_j^{\dagger}$ are defined in [12, (4.2)]. With the exception of replacing $n \mapsto j$ and $k \mapsto n$, we have preserved the notation for these constants from [12].

Remark 1.9. Our results apply to any $n \geq 2$, as the quantum modular properties in the case $n = 1$ readily follow from existing results. Namely, proceeding as in the proof of Theorem 3.2, one may determine a suitable quantum set for the normalized rank generating function in [6, Theorem 1.1]. Using [6, Theorem 1.1], a short calculation shows that the error to modularity (with respect to the nontrivial generator of Γ_c) is a multiple of

$$\int_x^{i\infty} \frac{\Theta(\frac{a}{c}; \ell_c \rho)}{\sqrt{-i(\tau + \rho)}} d\rho$$

for some $x \in \mathbb{Q}$. When viewed as a function of τ in a subset of $\mathbb{R}$, this integral is analytic (e.g., see [15, 19]).

One could also establish the quantum properties of a non-normalized version of R_1 by rewriting it in terms of the Appell-Lerch sum A_3, and proceeding as in the proof of Theorem 1.7. In this case, $R_1(\zeta_1; q)$ (where $\zeta_1 = e(\alpha_1/\beta_1)$) converges on the quantum set Q_{ζ_1}, where this set is defined by letting $n = 1$ in (1.13).

The interested reader may also wish to consult [9] for general results on quantum properties associated to mock modular forms.

Remark 1.10. In a forthcoming joint work [11], we extend Theorem 1.7 to hold for the more general vectors of roots of unity considered in [12], i.e., those with repeated entries. Allowing repeated roots of unity introduces additional singularities, and the modular completion of R_n is significantly more complicated. This precludes us from proving the more general case in the same way as the restricted case we address here.

2 Preliminaries

2.1 Modular, Mock Modular, and Jacobi Forms

A special ordinary modular form we require is Dedekind's η-function, defined in (1.1). This function is well known to satisfy the following transformation law [17].

Lemma 2.1. *For* $\gamma = \left(\begin{smallmatrix} a & b \\ c & d \end{smallmatrix} \right) \in \mathrm{SL}_2(\mathbb{Z})$, *we have that*

$$\eta(\gamma\tau) = \chi_\gamma (c\tau + d)^{\frac{1}{2}} \eta(\tau),$$

where

$$\chi_\gamma := \begin{cases} e\left(\frac{b}{24}\right), & \text{if } c = 0, d = 1, \\ \sqrt{-i}\, \omega_{d,c}^{-1} e\left(\frac{a+d}{24c}\right), & \text{if } c > 0, \end{cases}$$

with $\omega_{d,c} := e(\frac{1}{2}s(d,c))$. Here the Dedekind sum $s(m,t)$ is given for integers m and t by

$$s(m,t) := \sum_{j \bmod t} \left(\!\!\left(\frac{j}{t}\right)\!\!\right) \left(\!\!\left(\frac{mj}{t}\right)\!\!\right),$$

where $((x)) := x - \lfloor x \rfloor - 1/2$ if $x \in \mathbb{R} \setminus \mathbb{Z}$, and $((x)) := 0$ if $x \in \mathbb{Z}$.

The following gives a useful expression for χ_γ (see [14, Ch. 4, Thm. 2]):

$$\chi_\gamma = \begin{cases} \left(\frac{d}{|c|}\right) e\left(\frac{1}{24}\left((a+d)c - bd(c^2-1) - 3c\right)\right) & \text{if } c \equiv 1 \pmod 2, \\ \left(\frac{c}{d}\right) e\left(\frac{1}{24}\left((a+d)c - bd(c^2-1) + 3d - 3 - 3cd\right)\right) & \text{if } d \equiv 1 \pmod 2, \end{cases}$$

(2.1)

where $\left(\frac{\alpha}{\beta}\right)$ is the generalized Legendre symbol.

We require two additional functions, namely the Jacobi theta function $\vartheta(u; \tau)$, an ordinary Jacobi form, and a nonholomorphic modular-like function $R(u; \tau)$ used by Zwegers in [20]. In what follows, we will also need certain transformation properties of these functions.

Proposition 2.2. *For $u \in \mathbb{C}$ and $\tau \in \mathbb{H}$, define*

$$\vartheta(u; \tau) := \sum_{v \in \frac{1}{2} + \mathbb{Z}} e^{\pi i v^2 \tau + 2\pi i v \left(u + \frac{1}{2}\right)}.$$

(2.2)

Then ϑ satisfies

(1) $\vartheta(u+1; \tau) = -\vartheta(u; \tau)$,

(2) $\vartheta(u+\tau; \tau) = -e^{-\pi i \tau - 2\pi i u} \vartheta(u; \tau)$,

(3) $\vartheta(u; \tau) =$

$$-i e^{\pi i \tau / 4} e^{-\pi i u} \prod_{m=1}^{\infty} (1 - e^{2\pi i m \tau})(1 - e^{2\pi i u} e^{2\pi i \tau (m-1)})(1 - e^{-2\pi i u} e^{2\pi i m \tau}).$$

The nonholomorphic function $R(u; \tau)$ is defined in [20] by

$$R(u; \tau) := \sum_{v \in \frac{1}{2} + \mathbb{Z}} \left\{ \operatorname{sgn}(v) - E\left(\left(v + \frac{\operatorname{Im}(u)}{\operatorname{Im}(\tau)}\right) \sqrt{2 \operatorname{Im}(\tau)}\right) \right\} (-1)^{v - \frac{1}{2}} e^{-\pi i v^2 \tau - 2\pi i v u},$$

where

$$E(z) := 2 \int_0^z e^{-\pi t^2} dt.$$

The function R transforms like a (nonholomorphic) mock Jacobi form as follows.

Proposition 2.3 (Propositions 1.9 and 1.10 [20]). *The function R satisfies the following transformation properties:*

(1) $R(u + 1; \tau) = -R(u; \tau)$,

(2) $R(u; \tau) + e^{-2\pi i u - \pi i \tau} R(u + \tau; \tau) = 2e^{-\pi i u - \pi i \tau/4}$,

(3) $R(u; \tau) = R(-u; \tau)$,

(4) $R(u; \tau + 1) = e^{-\frac{\pi i}{4}} R(u; \tau)$,

(5) $\frac{1}{\sqrt{-i\tau}} e^{\pi i u^2/\tau} R\left(\frac{u}{\tau}; -\frac{1}{\tau}\right) + R(u; \tau) = h(u; \tau)$, *where the Mordell integral is defined by*

$$h(u; \tau) := \int_{\mathbb{R}} \frac{e^{\pi i \tau t^2 - 2\pi u t}}{\cosh \pi t} dt. \tag{2.3}$$

Using the functions ϑ and R, Zwegers defined the nonholomorphic function

$$\mathcal{R}_3(u, v; \tau) := \frac{i}{2} \sum_{j=0}^{2} e^{2\pi i j u} \vartheta(v + j\tau + 1; 3\tau) R(3u - v - j\tau - 1; 3\tau) \tag{2.4}$$

$$= \frac{i}{2} \sum_{j=0}^{2} e^{2\pi i j u} \vartheta(v + j\tau; 3\tau) R(3u - v - j\tau; 3\tau),$$

where the equality of the two expressions in (2.4) is justified by Proposition 2.2 and Proposition 2.3. This function is used to complete the level three Appell function (see [21] or [4])

$$A_3(u, v; \tau) := e^{3\pi i u} \sum_{n \in \mathbb{Z}} \frac{(-1)^n q^{3n(n+1)/2} e^{2\pi i n v}}{1 - e^{2\pi i u} q^n},$$

where $u, v \in \mathbb{C}$, as

$$\widehat{A}_3(u, v; \tau) := A_3(u, v; \tau) + \mathcal{R}_3(u, v; \tau).$$

This completed function transforms like a (nonholomorphic) Jacobi form, and in particular satisfies the following elliptic transformation.

Theoerm 2.4 ([21, Theorem 2.2]). *For $n_1, n_2, m_1, m_2 \in \mathbb{Z}$, the completed level 3 Appell function $\widehat{A}_3$ satisfies*

$$\widehat{A}_3(u + n_1\tau + m_1, v + n_2\tau + m_2; \tau)$$
$$= (-1)^{n_1+m_1} e^{2\pi i(u(3n_1 - n_2) - vn_1)} q^{3n_1^2/2 - n_1 n_2} \widehat{A}_3(u, v; \tau).$$

The following relationship between the Appell series A_3 and the combinatorial series R_n is proved in [12].

Proposition ([12, Proposition 4.2]). *Under the hypotheses given above on ζ_n, we have that*

$$R_n(\zeta_n; q) = \frac{1}{(q)_\infty} \sum_{j=1}^{n} \left(\zeta_{2\beta_j}^{-3\alpha_j} - \zeta_{2\beta_j}^{-\alpha_j} \right) \frac{A_3\left(\frac{\alpha_j}{\beta_j}, -2\tau; \tau \right)}{\Pi_j^\dagger(\alpha_n)}.$$

We also note that

$$\widehat{\mathcal{A}}_n(\tau) = \frac{1}{\eta(\tau)} \sum_{j=1}^{n} (\zeta_{2\beta_j}^{-3\alpha_j} - \zeta_{2\beta_j}^{-\alpha_j}) \frac{\widehat{A}_3\left(\frac{\alpha_j}{\beta_j}, -2\tau; \tau \right)}{\Pi_j^\dagger(\alpha_n)}.$$

In addition to working with the Appell sum $\widehat{A}_3$, we also make use of additional properties of the functions h and R. In particular, Zwegers also showed how under certain hypotheses, these functions can be written in terms of integrals involving the weight 3/2 modular forms $g_{a,b}(\tau)$, defined for $a, b \in \mathbb{R}$ and $\tau \in \mathbb{H}$ by

$$g_{a,b}(\tau) := \sum_{v \in a+\mathbb{Z}} v e^{\pi i v^2 \tau + 2\pi i v b}. \tag{2.5}$$

We will make use of the following results.

Proposition 2.5 ([20, Proposition 1.15 (1), (2), (4), (5)]). *The function $g_{a,b}$ satisfies:*

(1) $g_{a+1,b}(\tau) = g_{a,b}(\tau)$,

(2) $g_{a,b+1}(\tau) = e^{2\pi i a} g_{a,b}(\tau)$,

(3) $g_{a,b}(\tau + 1) = e^{-\pi i a(a+1)} g_{a,a+b+\frac{1}{2}}(\tau)$,

(4) $g_{a,b}(-\frac{1}{\tau}) = i e^{2\pi i a b} (-i\tau)^{\frac{3}{2}} g_{b,-a}(\tau)$.

Theoerm 2.6 ([20, Theorem 1.16 (2)]). *Let $\tau \in \mathbb{H}$. For $a, b \in (-\frac{1}{2}, \frac{1}{2})$, we have*

$$h(a\tau - b; \tau) = -e\left(\frac{a^2\tau}{2} - a(b + \frac{1}{2}) \right) \int_0^{i\infty} \frac{g_{a+\frac{1}{2}, b+\frac{1}{2}}(\rho)}{\sqrt{-i(\rho + \tau)}} d\rho.$$

3 The Quantum Set

We call a subset $S \subseteq \mathbb{Q}$ a *quantum set* for a function F with respect to the group $G \subseteq \mathrm{SL}_2(\mathbb{Z})$ if both $F(x)$ and $F(Mx)$ exist (are non-singular) for all $x \in S$ and $M \in G$.

In this section, we will show that Q_{ζ_n} as defined in (1.13) is a quantum set for $\mathcal{A}_n$ with respect to the group Γ_{ζ_n}. Recall that Q_{ζ_n} is defined as

$$
Q_{\zeta_n} := \left\{ \frac{h}{k} \in \mathbb{Q} \;\middle|\; \begin{array}{l} h \in \mathbb{Z}, k \in \mathbb{N}, \gcd(h, k) = 1, \; \beta_j \nmid k \; \forall \, 1 \le j \le n, \\[4pt] \left| \dfrac{\alpha_j}{\beta_j} k - \left[\dfrac{\alpha_j}{\beta_j} k \right] \right| > \dfrac{1}{6} \; \forall \, 1 \le j \le n \end{array} \right\},
$$

where $[x]$ is the closest integer to x (see Remark 1.6).

Moreover, recall that the "holomorphic part" we consider (see § 1.3) is $\mathcal{A}_n(\tau) = q^{-\frac{1}{24}} R_n(\boldsymbol{\zeta_n}; q)$. To show that Q_{ζ_n} is a quantum set for $\mathcal{A}_n(\tau)$, we must first show that the multi-sum defining $R_n(\boldsymbol{\zeta_n}; \zeta_k^h)$ converges for $\frac{h}{k} \in Q_{\zeta_n}$. In what follows, as in the definition of Q_{ζ_n}, we take $h \in \mathbb{Z}, k \in \mathbb{N}$ such that $\gcd(h, k) = 1$.

We start by addressing the restriction that for $\frac{h}{k} \in Q_{\zeta_n}$, $\beta_j \nmid k$ for all $1 \le j \le n$.

Lemma 3.1. *For $\frac{h}{k} \in \mathbb{Q}$, all summands of $R_n(\boldsymbol{\zeta_n}; \zeta_k^h)$ are finite if and only if $\beta_j \nmid k$ for all $1 \le j \le n$.*

Proof. Examining the multi-sum $R_n(\boldsymbol{\zeta_n}; \zeta_k^h)$, we see that all terms are a power of ζ_k^h divided by a product of factors of the form $1 - \zeta_{\beta_j}^{\pm \alpha_j} \zeta_k^{hm}$ for some integer $m \ge 1$. Therefore, to have each summand be finite, it is enough to ensure that $1 - \zeta_{\beta_j}^{\pm \alpha_j} \zeta_k^{hm} \ne 0$ for all $m \ge 1$ and for all $1 \le j \le n$. For ease of notation in this proof, we will omit the subscripts for α_j and β_j.

If $1 - \zeta_{\beta}^{\pm \alpha} \zeta_k^{hm} = 0$ for some $m \in \mathbb{N}$, we have that

$$
\pm \frac{\alpha}{\beta} + \frac{hm}{k} \in \mathbb{Z}.
$$

Let $K = \mathrm{lcm}(\beta, k) = \beta \beta' = kk'$. Then $\pm \frac{\alpha}{\beta} + \frac{hm}{k} \notin \mathbb{Z}$ is the same as $\pm \alpha \beta' + hmk' \notin K\mathbb{Z}$. Since $K = kk'$, if $k' \nmid \alpha \beta'$, this is always true and we do not have a singularity.

However, since $K = \beta \beta' = kk'$, if $k' | \alpha \beta'$, then $\frac{\beta \beta'}{k} | \alpha \beta'$. This implies that $\beta | \alpha k$ and that $\beta | k$ since $\gcd(\alpha, \beta) = 1$.

Therefore, if $\beta \nmid k$, it is always the case that $k' \nmid \alpha \beta'$, so for all $m \in \mathbb{N}$,

$$
\pm \frac{\alpha}{\beta} + \frac{hm}{k} \notin \mathbb{Z}.
$$

Now that we have shown that all summands in $R_n(\boldsymbol{\zeta_n}; \zeta_k^h)$ are finite for $\frac{h}{k} \in Q_{\zeta_n}$, we will show that the sum converges.

Theoerm 3.2. *For ζ_n as in (1.8), if $\frac{h}{k} \in Q_{\zeta_n}$, then $R_n(\zeta_n; \zeta_k^h)$ converges and can be evaluated as a finite sum. In particular, we have that:*

$$R_n(\zeta_n; \zeta_k^h) = \prod_{j=1}^{n} \frac{1}{1 - ((1 - x_j^k)(1 - x_j^{-k}))^{-1}}$$

$$\times \sum_{\substack{0 < m_1 \le k \\ 0 \le m_2, \dots, m_n < k}} \frac{\zeta_k^{h[(m_1+m_2+\cdots+m_n)^2+(m_1+\cdots+m_{n-1})+(m_1+\cdots+m_{n-2})+\cdots+m_1]}}{(x_1\zeta_k^h; \zeta_k^h)_{m_1} \left(\frac{\zeta_k^h}{x_1}; \zeta_k^h\right)_{m_1} (x_2\zeta_k^{hm_1}; \zeta_k^h)_{m_2+1} \left(\frac{\zeta_k^{hm_1}}{x_2}; \zeta_k^h\right)_{m_2+1}}$$

$$\times \frac{1}{(x_3\zeta_k^{h(m_1+m_2)}; \zeta_k^h)_{m_3+1}\left(\frac{\zeta_k^{h(m_1+m_2)}}{x_3}; \zeta_k^h\right)_{m_3+1} \cdots (x_n\zeta_k^{h(m_1+\cdots+m_{n-1})}; \zeta_k^h)_{m_n+1}\left(\frac{\zeta_k^{h(m_1+\cdots+m_{n-1})}}{x_n}; \zeta_k^h\right)_{m_n+1}},$$

$$(3.1)$$

where $\zeta_n = (x_1, x_2, \dots, x_n)$.

Proof of Theorem 3.2. We start by taking $\frac{h}{k} \in Q_{\zeta_n}$, and write $\zeta = \zeta_k^h$. For ease of notation, we will use x_j to denote the j-th component in ζ_n, so $x_j = e^{2\pi i\alpha_j/\beta_j}$. Furthermore, for clarity of argument, we will carry out the proof in the case of $n = 2$, with comments throughout about how the proof follows for $n > 2$. We have that

$$R_2((x_1, x_2); \zeta) = \sum_{\substack{m_1 > 0 \\ m_2 \ge 0}} \frac{\zeta^{(m_1+m_2)^2+m_1}}{(x_1\zeta; \zeta)_{m_1}(x_1^{-1}\zeta; \zeta)_{m_1}(x_2\zeta^{m_1}; \zeta)_{m_2+1}(x_2^{-1}\zeta^{m_1}; \zeta)_{m_2+1}}$$

$$= \sum_{M_1, M_2 \ge 0} \frac{1}{(1 - x_1^k)^{M_1}(1 - x_1^{-k})^{M_1}(1 - x_2^k)^{M_2}(1 - x_2^{-k})^{M_2}}$$

$$(3.2)$$

$$\times \sum_{\substack{0 < s_1 \le k \\ 0 \le s_2 < k}} \frac{\zeta^{(s_1+s_2)^2+s_1}}{(x_1\zeta; \zeta)_{s_1}(x_1^{-1}\zeta; \zeta)_{s_1}(x_2\zeta^{s_1}; \zeta)_{s_2+1}(x_2^{-1}\zeta^{s_1}; \zeta)_{s_2+1}},$$

$$(3.3)$$

where we have let $m_j = s_j + M_jk$ for $0 < s_1 \le k$, $0 \le s_2 < k$, and $M_j \in \mathbb{N}_0$, and have used the fact that

$$(x\zeta^r; \zeta)_{s+Mk} = (1 - x^k)^M (x\zeta^r; \zeta)_s,$$

which holds for any $M, r, s \in \mathbb{N}_0$. (We note that for $n > 2$, we proceed as above, additionally taking $0 \le s_j \le k - 1$ for $j > 2$.) The second sum in (3.3) is a finite sum, as desired. For the first sum in (3.2) we notice that we in fact have the product of two geometric series, each of the form

$$\sum_{M_j \geq 0} \left(\frac{1}{(1 - x_j^k)(1 - x_j^{-k})} \right)^{M_j}.$$

By definition, we have $x_j = \cos \theta_j + i \sin \theta_j$ where $\theta_j = \frac{2\pi \alpha_j}{\beta_j}$. Therefore, this sum converges if and only if

$$|1 - x_j^k||1 - x_j^{-k}| = 2 - 2\cos(k\theta_j) > 1 \iff \cos(k\theta_j) < \frac{1}{2}.$$

For $\cos(k\theta_j) < \frac{1}{2}$, it must be that $k\theta_j = r + 2\pi M$ where $-\pi < r \leq \pi$, $|r| > \frac{\pi}{6}$, and $M \in \mathbb{Z}$. This is equivalent to saying

$$\left| \frac{\alpha_j}{\beta_j} k - \left[\frac{\alpha_j}{\beta_j} k \right] \right| > \frac{1}{6} \ \forall \ 1 \leq j \leq n,$$

as in the definition of Q_{ζ_n} in (1.13). Therefore, we see that for $\frac{h}{k} \in Q_{\zeta_n}$, $R_2((x_1, x_2); \zeta)$ converges to the claimed expression in (3.1).

We note that by Abel's theorem, having shown convergence of $R_2((x_1, x_2); \zeta)$, we have that $R_2((x_1, x_2); q)$ converges to $R_2((x_1, x_2); \zeta)$ as $q \to \zeta$ radially within the unit disc.

As noted, the above argument extends to $n > 2$. Letting $m_j = s_j + M_j k$ with $0 < s_1 \leq k$ and $0 \leq s_j < k$ for $j \geq 2$, rewriting as in (3.1), and then summing the resulting geometric series gives the desired exact formula for $R_n(\zeta_n; \zeta)$.

To complete the argument that Q_{ζ_n} is a quantum set for $R_n(\zeta_n; \zeta)$ with respect to Γ_{ζ_n}, it remains to be seen that $R_n(\zeta_n; \xi)$ converges, where $\xi = e^{2\pi i \gamma(\frac{h}{k})}$ for $\frac{h}{k} \in Q_{\zeta_n}$ and $\gamma \in \Gamma_{\zeta_n}$, defined in (1.15). For the ease of the reader, we recall from (1.14) and (1.15) that

$$\Gamma_{\zeta_n} := \left\langle \begin{pmatrix} 1 & 1 \\ 0 & 1 \end{pmatrix}, \begin{pmatrix} 1 & 0 \\ \ell & 1 \end{pmatrix} \right\rangle,$$

where

$$\ell = \ell_\beta := \begin{cases} 6 \, [\mathrm{lcm}(\beta_1, \ldots, \beta_k)]^2 & \text{if } \forall j, 3 \nmid \beta_j \\ 2 \, [\mathrm{lcm}(\beta_1, \ldots, \beta_k)]^2 & \text{if } \exists j, 3 \mid \beta_j. \end{cases}$$

The convergence of $R_n(\zeta_n; \xi)$ is a direct consequence of the following lemma.

Lemma 3.3. *The set Q_{ζ_n} is closed under the action of Γ_{ζ_n}.*

Proof. Since Γ_{ζ_n} is given as a set with two generators, it is enough to show that Q_{ζ_n} is closed under action of each of those generators.

Let $\frac{h}{k} \in Q_{\zeta_n}$. Then $\begin{pmatrix} 1 & 1 \\ 0 & 1 \end{pmatrix} \frac{h}{k} = \frac{h+k}{k}$. Note that $\gcd(h+k, k) = \gcd(h, k) = 1$ and we already know that k satisfies the conditions in the definition of Q_{ζ_n}. Therefore, $\begin{pmatrix} 1 & 1 \\ 0 & 1 \end{pmatrix} \frac{h}{k} \in Q_{\zeta_n}$.

Under the action of $\begin{pmatrix} 1 & 0 \\ \ell & 1 \end{pmatrix}$, we have

$$\begin{pmatrix} 1 & 0 \\ \ell & 1 \end{pmatrix} \frac{h}{k} = \frac{h}{h\ell + k}.$$

We first note that $\gcd(h, h\ell + k) = \gcd(h, k) = 1$, and $\beta_j \nmid (h\ell + k)$ as $\beta_j | \ell$ and $\beta_j \nmid k$. It remains to check that

$$\left| \frac{\alpha_j}{\beta_j}(h\ell + k) - \left[\frac{\alpha_j}{\beta_j}(h\ell + k) \right] \right| > \frac{1}{6} \ \forall \ 1 \leq j \leq n.$$

We have that

$$\left| \frac{\alpha_j}{\beta_j}(h\ell + k) - \left[\frac{\alpha_j}{\beta_j}(h\ell + k) \right] \right| = \left| \frac{\alpha_j h\ell}{\beta_j} + \frac{\alpha_j}{\beta_j}k - \left[\frac{\alpha_j h\ell}{\beta_j} + \frac{\alpha_j}{\beta_j}k \right] \right|$$

$$= \left| \frac{\alpha_j}{\beta_j}k - \left[\frac{\alpha_j}{\beta_j}k \right] \right| > \frac{1}{6}, \tag{3.4}$$

where we can simplify as in (3.4) since, by definition of ℓ, $\frac{\alpha_j \ell}{\beta_j} \in \mathbb{Z}$. Thus, Q_{ζ_n} is closed under the action of Γ_{ζ_n}.

4 Proof of Theorem 1.7

We now prove Theorem 1.7. Our first goal is to establish that $H_{n,\gamma}$ is analytic in x on $\mathbb{R} - \{\frac{-c}{d}\}$ for all $x \in Q_{\zeta_n}$ and $\gamma = \begin{pmatrix} a & b \\ c & d \end{pmatrix} \in \Gamma_{\zeta_n}$. As shown in Section 3, we have that $\mathcal{A}_n(x)$ and $\mathcal{A}_n(\gamma x)$ are defined for all $x \in Q_{\zeta_n}$ and $\gamma \in \Gamma_{\zeta_n}$. Note that it suffices to consider only the generators S_ℓ and T of Γ_{ζ_n}, since

$$H_{n,\gamma\gamma'}(\tau) = H_{n,\gamma'}(\tau) + \chi_{\gamma'}(C\tau + D)^{-\frac{1}{2}} H_{n,\gamma}(\gamma'\tau)$$

for $\gamma = \begin{pmatrix} a & b \\ c & d \end{pmatrix}$ and $\gamma' = \begin{pmatrix} A & B \\ C & D \end{pmatrix}$.

First, consider $\gamma = T$. Then by definition, $\chi_T = \zeta_{24}$, and so $H_{n,T}(x) = \mathcal{A}_n(x) - \zeta_{24}\mathcal{A}_n(x + 1)$. When we map $x \mapsto x + 1$, $q = e^{2\pi i x}$ remains invariant. Then since the definition of $R_n(x)$ in (1.6) can be expressed as a series only involving integer powers of q, it is also invariant. Thus

$$\mathcal{A}_n(x + 1) = e^{\frac{-2\pi i(x+1)}{24}} R_n(x) = \zeta_{24}^{-1} \mathcal{A}_n(x),$$

and so $H_{n,T}(x) = 0$.

We now consider the case $\gamma = S_\ell$. In this case using (2.1) we calculate that $\chi_{S_\ell} = \zeta_{24}^{-\ell}$. Thus,

$$H_{n,S_\ell}(x) = A_n(x) - \zeta_{24}^{-\ell}(\ell x + 1)^{-\frac{1}{2}} A_n(S_\ell x).$$

From the modularity of $\widehat{A}_n$ we have that $\widehat{A}_n(x) = \zeta_{24}^{-\ell}(\ell x + 1)^{-\frac{1}{2}} \widehat{A}_n(S_\ell x)$. Thus (1.9) and (1.12) give that

$$H_{n,S_\ell}(x) = -A_n^-(x) + \zeta_{24}^{-\ell}(\ell x + 1)^{-\frac{1}{2}} A_n^-(S_\ell x), \tag{4.1}$$

where A_n^- is defined in (1.10).

Using the Jacobi triple product identity from Proposition 2.2 item (3), we can simplify the theta functions to get that $\vartheta(-2\tau; 3\tau) = iq^{-\frac{2}{3}}\eta(\tau)$, $\vartheta(-\tau; 3\tau) = iq^{-\frac{1}{6}}\eta(\tau)$, and $\vartheta(0; 3\tau) = 0$. Thus,

$$\mathcal{R}_3\left(\frac{\alpha_j}{\beta_j}, -2\tau; \tau\right) = -\frac{1}{2}q^{-\frac{2}{3}}\eta(\tau)\sum_{\delta=0}^{1} e\left(\frac{\alpha_j}{\beta_j}\delta\right) q^{\frac{\delta}{2}} R\left(\frac{3\alpha_j}{\beta_j} + (2-\delta)\tau; 3\tau\right).$$

Using Proposition 2.3 item (2), we can rewrite

$$R\left(\frac{3\alpha_j}{\beta_j} + 2\tau; 3\tau\right) = 2e\left(\frac{3\alpha_j}{2\beta_j}\right) q^{\frac{5}{8}} - e\left(\frac{3\alpha_j}{\beta_j}\right) q^{\frac{1}{2}} R\left(\frac{3\alpha_j}{\beta_j} - \tau; 3\tau\right),$$

so that

$$\sum_{\delta=0}^{1} e\left(\frac{\alpha_j}{\beta_j}\delta\right) q^{\frac{\delta}{2}} R\left(\frac{3\alpha_j}{\beta_j} + (2-\delta)\tau; 3\tau\right) =$$
$$2e\left(\frac{3\alpha_j}{2\beta_j}\right) q^{\frac{5}{8}} + e\left(\frac{2\alpha_j}{\beta_j}\right) q^{\frac{1}{2}} \sum_{\pm} \pm e\left(\mp\frac{\alpha_j}{\beta_j}\right) R\left(\frac{3\alpha_j}{\beta_j} \pm \tau; 3\tau\right).$$

Thus we see that

$$A_n^-(\tau) = -\frac{1}{2}\sum_{j=1}^{n} \frac{(\zeta_{2\beta_j}^{-3\alpha_j} - \zeta_{2\beta_j}^{-\alpha_j})}{\Pi_j^\dagger(\alpha_k)} e\left(\frac{2\alpha_j}{\beta_j}\right) q^{-\frac{1}{6}} \sum_{\pm} \pm e\left(\mp\frac{\alpha_j}{\beta_j}\right) R\left(\frac{3\alpha_j}{\beta_j} \pm \tau; 3\tau\right)$$
$$-q^{-\frac{1}{24}}\sum_{j=1}^{n} \frac{(\zeta_{2\beta_j}^{-3\alpha_j} - \zeta_{2\beta_j}^{-\alpha_j})}{\Pi_j^\dagger(\alpha_k)} e\left(\frac{3\alpha_j}{2\beta_j}\right). \tag{4.2}$$

Now to compute $A_n^-(S_\ell \tau)$ we first define

$$F_{\alpha,\beta}(\tau) := q^{-\frac{1}{6}} \sum_{\pm} \pm e\left(\mp\frac{\alpha}{\beta}\right) R\left(\frac{3\alpha}{\beta} \pm \tau; 3\tau\right).$$

Then by (4.1) and (4.2) we can write

$$H_{n,S_\ell}(\tau) = \frac{1}{2} \sum_{j=1}^{n} \frac{(\zeta_{2\beta_j}^{-3\alpha_j} - \zeta_{2\beta_j}^{-\alpha_j})}{\Pi_j^\dagger(\alpha_k)} e\left(\frac{2\alpha_j}{\beta_j}\right) \left[F_{\alpha_j,\beta_j}(\tau) - \zeta_{24}^{-\ell}(\ell\tau+1)^{-\frac{1}{2}} F_{\alpha_j,\beta_j}(S_\ell\tau)\right]$$

$$+ \sum_{j=1}^{n} \frac{(\zeta_{2\beta_j}^{-3\alpha_j} - \zeta_{2\beta_j}^{-\alpha_j})}{\Pi_j^\dagger(\alpha_k)} (\ell\tau+1)^{-\frac{1}{2}} \zeta_{24}^{-\ell} \mathcal{E}_1\left(\frac{\alpha_j}{\beta_j}, \ell; \tau\right),$$

where

$$\mathcal{E}_1\left(\frac{\alpha}{\beta}, \ell; \tau\right) := (\ell\tau+1)^{\frac{1}{2}} \zeta_{24}^{\ell} q^{-\frac{1}{24}} e\left(\frac{3}{2}\frac{\alpha}{\beta}\right) - e\left(\frac{-S_\ell\tau}{24}\right) e\left(\frac{3}{2}\frac{\alpha}{\beta}\right). \tag{4.3}$$

Thus in order to prove that $H_{n,S_\ell}(x)$ is analytic on $\mathbb{R} - \{\frac{-1}{\ell}\}$ it suffices to show that for each $1 \le j \le n$,

$$G_{\alpha_j,\beta_j}(\tau) := F_{\alpha_j,\beta_j}(\tau) - \zeta_{24}^{-\ell}(\ell\tau+1)^{-\frac{1}{2}} F_{\alpha_j,\beta_j}(S_\ell\tau)$$

is analytic on $\mathbb{R} - \{\frac{-1}{\ell}\}$. We establish this in Proposition 4.1 below.

Proposition 4.1. *Fix* $1 \le j \le n$ *and set* $(\alpha, \beta) := (\alpha_j, \beta_j)$. *With notation and hypotheses as above, we have that*

$$G_{\alpha,\beta}(\tau) = \sqrt{3} \sum_{\pm} \mp e\left(\mp\frac{1}{6}\right) \int_{\frac{1}{\ell}}^{i\infty} \frac{g_{\pm\frac{1}{3}+\frac{1}{2},\frac{1}{2}-3\frac{\alpha}{\beta}}(3\rho)}{\sqrt{-i(\rho+\tau)}} d\rho,$$

which is analytic on $\mathbb{R} - \left\{\frac{-1}{\ell}\right\}$.

Proof. Fix $1 \le j \le n$ and set $(\alpha, \beta) := (\alpha_j, \beta_j)$. Define $m := \left[\frac{3\alpha}{\beta}\right] \in \mathbb{Z}$, $r \in (-\frac{1}{2}, \frac{1}{2})$ so that $\frac{3\alpha}{\beta} = m + r$. We note that $r \ne \pm\frac{1}{2}$ since $\beta \ne 2$. Using Proposition 2.3 (1), we have that

$$F_{\alpha,\beta}(\tau) = q^{-\frac{1}{6}} \sum_{\pm} \pm e\left(\frac{\mp r}{3}\right) e\left(\frac{\mp m}{3}\right) (-1)^m R\left(\pm\tau + r; 3\tau\right). \tag{4.4}$$

Letting $\tau_\ell := -\frac{1}{\tau} - \ell$ we have $S_\ell\tau = \frac{-1}{\tau_\ell}$. Using Proposition 2.3 (5) with $u = \frac{r}{3}\tau_\ell \mp \frac{1}{3}$ and $\tau \mapsto \frac{\tau_\ell}{3}$ we see that

$$R\left(r \mp \frac{1}{\tau_\ell}; \frac{-3}{\tau_\ell}\right) =$$

$$\sqrt{\frac{-i\tau_\ell}{3}} \cdot e\left(-\frac{1}{2}\left(\frac{r\tau_\ell}{3} \mp \frac{1}{3}\right)^2\left(\frac{3}{\tau_\ell}\right)\right)\left[h\left(\frac{r\tau_\ell}{3} \mp \frac{1}{3}; \frac{\tau_\ell}{3}\right) - R\left(\frac{r\tau_\ell}{3} \mp \frac{1}{3}; \frac{\tau_\ell}{3}\right)\right].$$

$$(4.5)$$

Using Proposition 2.3 parts (1) and (4) we see that $R\left(\frac{r\tau_\ell}{3} \mp \frac{1}{3}; \frac{\tau_\ell}{3}\right) = \zeta_{24}^\ell R\left(\frac{-r}{3\tau} \mp \frac{1}{3}; \frac{-1}{3\tau}\right)$. Then using Proposition 2.3 (5) with $u = \mp\tau - r$ and $\tau \mapsto 3\tau$ we obtain that

$$R\left(\frac{r\tau_\ell}{3} \mp \frac{1}{3}; \frac{\tau_\ell}{3}\right) = \zeta_{24}^\ell \sqrt{-i(3\tau)} \cdot e\left(\frac{-(\mp\tau - r)^2}{6\tau}\right)[h(\mp\tau - r; 3\tau) - R(\mp\tau - r; 3\tau)],$$

which together with (4.4) and (4.5) gives

$$F_{\alpha,\beta}(S_\ell\tau) =$$

$$e\left(\frac{1}{6\tau_\ell}\right)\sum_\pm \pm e\left(\frac{\mp r}{3}\right)e\left(\frac{\mp m}{3}\right)(-1)^m\sqrt{\frac{-i\tau_\ell}{3}} \cdot e\left(-\frac{1}{2}\left(\frac{r\tau_\ell}{3} \mp \frac{1}{3}\right)^2\left(\frac{3}{\tau_\ell}\right)\right) \cdot$$

$$\left[h\left(\frac{r\tau_\ell}{3} \mp \frac{1}{3}; \frac{\tau_\ell}{3}\right) - \zeta_{24}^\ell\sqrt{-i(3\tau)} \cdot e\left(\frac{-(\mp\tau-r)^2}{6\tau}\right)[h(\mp\tau-r; 3\tau) - R(\mp\tau - r; 3\tau)]\right].$$

By the definition of r and ℓ we have that $\frac{r^2\ell}{6} \in \mathbb{Z}$. Simplifying thus gives that

$$F_{\alpha,\beta}(S_\ell\tau) = \sum_\pm \pm(-1)^m e\left(\frac{\mp m}{3}\right)e\left(\frac{r^2}{6\tau}\right)\sqrt{\frac{-i\tau_\ell}{3}}h\left(\frac{r\tau_\ell}{3} \mp \frac{1}{3}; \frac{\tau_\ell}{3}\right)$$

$$- \sum_\pm \pm(-1)^m e\left(\frac{\mp m}{3}\right)e\left(\frac{\mp r}{3}\right)q^{-\frac{1}{6}}\zeta_{24}^\ell(\ell\tau + 1)^{\frac{1}{2}} \cdot h(\mp\tau - r; 3\tau)$$

$$+ \sum_\pm \pm(-1)^m e\left(\frac{\mp m}{3}\right)e\left(\frac{\mp r}{3}\right)q^{-\frac{1}{6}}\zeta_{24}^\ell(\ell\tau + 1)^{\frac{1}{2}} \cdot R(\mp\tau - r; 3\tau),$$

and so using Proposition 2.3 (3) and the fact that $h(u; \tau) = h(-u; \tau)$ which comes directly from the definition of h in (2.3), we see that

$$G_{\alpha,\beta}(\tau) = q^{-\frac{1}{6}}\sum_\pm \pm(-1)^m e\left(\frac{\mp m}{3}\right)e\left(\frac{\mp r}{3}\right)h(\pm\tau + r; 3\tau)$$

$$- \sum_\pm \pm(-1)^m e\left(\frac{\mp m}{3}\right)e\left(\frac{r^2}{6\tau}\right)\zeta_{24}^{-\ell}\sqrt{\frac{i}{3\tau}} \cdot h\left(\frac{r\tau_\ell}{3} \mp \frac{1}{3}; \frac{\tau_\ell}{3}\right).$$

We now use Theorem 2.6 to convert the h functions into integrals. Letting $a = \frac{\pm 1}{3}$, $b = -r$, and $\tau \mapsto 3\tau$ gives that

$$h(\pm\tau + r; 3\tau) = -q^{\frac{1}{6}}\zeta_6^{\mp 1}e\left(\frac{\pm r}{3}\right)\int_0^{i\infty}\frac{g_{\pm\frac{1}{3}+\frac{1}{2},\frac{1}{2}-r}(z)dz}{\sqrt{-i(z+3\tau)}}.$$

Letting $a = r$, $b = \frac{\pm 1}{3}$, and $\tau \mapsto \frac{\tau_\ell}{3}$ gives that

$$h\left(\frac{r\tau_\ell}{3}\mp\frac{1}{3};\frac{\tau_\ell}{3}\right) = -e\left(\frac{-r^2}{6\tau}\right)e\left(\frac{\mp r}{3}\right)e\left(\frac{-r}{2}\right)\int_0^{i\infty}\frac{g_{r+\frac{1}{2},\pm\frac{1}{3}+\frac{1}{2}}(z)dz}{\sqrt{-i\left(z+\frac{\tau_\ell}{3}\right)}}.$$

Thus

$$G_{\alpha,\beta}(\tau) = -\sum_{\pm}\pm\zeta_6^{\mp 1}(-1)^m e\left(\frac{\mp m}{3}\right)\int_0^{i\infty}\frac{g_{\pm\frac{1}{3}+\frac{1}{2},\frac{1}{2}-r}(z)dz}{\sqrt{-i(z+3\tau)}}$$

$$+ \sum_{\pm}\pm\zeta_{24}^{-\ell}(-1)^m e\left(\frac{\mp m}{3}\right)e\left(\frac{\mp r}{3}\right)e\left(\frac{-r}{2}\right)\sqrt{\frac{i}{3\tau}}\int_0^{i\infty}\frac{g_{r+\frac{1}{2},\pm\frac{1}{3}+\frac{1}{2}}(z)dz}{\sqrt{-i\left(z+\frac{\tau_\ell}{3}\right)}}.$$

By a simple change of variables (let $z = \frac{\ell}{3} - \frac{1}{z}$) we can write

$$\int_0^{i\infty}\frac{g_{r+\frac{1}{2},\pm\frac{1}{3}+\frac{1}{2}}(z)dz}{\sqrt{-i\left(z+\frac{\tau_\ell}{3}\right)}} = -\sqrt{-3\tau}\int_{\frac{3}{\ell}}^{0}\frac{g_{r+\frac{1}{2},\pm\frac{1}{3}+\frac{1}{2}}\left(\frac{\ell}{3}-\frac{1}{z}\right)dz}{z^{\frac{3}{2}}\sqrt{-i(z+3\tau)}}. \qquad (4.6)$$

Moreover, using Proposition 2.5 we can convert

$$g_{r+\frac{1}{2},\pm\frac{1}{3}+\frac{1}{2}}\left(\frac{\ell}{3}-\frac{1}{z}\right) = \zeta_{24}^\ell \cdot g_{r-\frac{1}{2},\pm\frac{1}{3}+\frac{1}{2}}\left(\frac{-1}{z}\right)$$

$$= -\zeta_{24}^\ell e\left(\frac{1}{8}\right)e\left(\frac{\mp 1}{6}\right)e\left(\frac{\pm r}{3}\right)e\left(\frac{r}{2}\right)z^{\frac{3}{2}}\cdot g_{\pm\frac{1}{3}+\frac{1}{2},\frac{1}{2}-r}(z). \qquad (4.7)$$

Thus by (4.6) and (4.7) we have that

$$G_{\alpha,\beta}(\tau) = -\sum_{\pm}\pm\zeta_6^{\mp 1}(-1)^m e\left(\frac{\mp m}{3}\right)\int_0^{i\infty}\frac{g_{\pm\frac{1}{3}+\frac{1}{2},\frac{1}{2}-r}(z)dz}{\sqrt{-i(z+3\tau)}}$$

$$- \sum_{\pm}\pm\zeta_6^{\mp 1}(-1)^m e\left(\frac{\mp m}{3}\right)\int_{\frac{3}{\ell}}^{0}\frac{g_{\pm\frac{1}{3}+\frac{1}{2},\frac{1}{2}-r}(z)dz}{\sqrt{-i(z+3\tau)}}$$

$$= -\sum_{\pm} \pm \zeta_6^{\mp 1} (-1)^m e\left(\frac{\mp m}{3}\right) \int_{\frac{3}{\ell}}^{i\infty} \frac{g_{\pm\frac{1}{3}+\frac{1}{2},\frac{1}{2}-r}(z)dz}{\sqrt{-i(z+3\tau)}}. \qquad (4.8)$$

To complete the proof, one can deduce from Proposition 2.5 (2) that for $m \in \mathbb{Z}$,

$$g_{a,b}(\tau) = e(ma)g_{a,b-m}(\tau).$$

Applying this to (4.8) with a direct calculation gives us

$$G_{\alpha,\beta}(\tau) = \sqrt{3}\sum_{\pm} \mp e\left(\mp\frac{1}{6}\right) \int_{\frac{1}{\ell}}^{i\infty} \frac{g_{\pm\frac{1}{3}+\frac{1}{2},\frac{1}{2}-3\frac{\alpha}{\beta}}(3z)}{\sqrt{-i(z+\tau)}}dz,$$

which is analytic on $\mathbb{R} - \{\frac{-1}{\ell}\}$ as desired.

5 Conclusion

We have proven that when we restrict to vectors $\boldsymbol{\zeta_n}$ which contain distinct roots of unity, the mock modular form $q^{-\frac{1}{24}}R_n(\boldsymbol{\zeta_n};q)$ is also a quantum modular form. To consider the more general case where we allow roots of unity in $\boldsymbol{\zeta_n}$ to repeat, the situation is significantly more complicated. In this setting, as shown in [12], the nonholomorphic completion of $q^{-\frac{1}{24}}R_n(\boldsymbol{\zeta_n};q)$ is not modular, but is instead a sum of two (nonholomorphic) modular forms of different weights. We will address this more general case in a forthcoming paper [11].

References

1. G. E. Andrews *Partitions, Durfee symbols, and the Atkin-Garvan moments of ranks*, Invent. Math. 169 (2007), 37–73.
2. A. O. L. Atkin and H. P. F. Swinnerton-Dyer, *Some properties of partitions*, Proc. London Math. Soc. 66 (1954), 84–106.
3. K. Bringmann, *On the explicit construction of higher deformations of partition statistics*, Duke Math. J., 144 (2008), 195-233.
4. K. Bringmann, A. Folsom, K. Ono, and L. Rolen, *Harmonic Maass forms and mock modular forms: theory and applications*, American Mathematical Society Colloquium Publications. American Mathematical Society, Providence, RI, 64 (2017).
5. K. Bringmann, F. Garvan, and K. Mahlburg, *Partition statistics and quasiweak Maass forms*, Int. Math. Res. Notices, 1 (2009), 63–97.
6. K. Bringmann and K. Ono, *Dyson's ranks and Maass forms*, Ann. of Math., 171 (2010), 419-449.
7. K. Bringmann and L. Rolen, *Radial limits of mock theta functions*, Res. Math. Sci. 2 (2015), Art. 17, 18 pp.
8. J. Bruiner and J. Funke, *On two geometric theta lifts*, Duke Math. J. 125 (2004), 45-90.

9. D. Choi, S. Lim, and R.C. Rhoades, *Mock modular forms and quantum modular forms,* Proc. Amer. Math. Soc. 144 (2016), no. 6, 2337–2349.

10. F. Dyson, *Some guesses in the theory of partitions,* Eureka (Cambridge) 8 (1944), 10–15.

11. A. Folsom, M-J. Jang, S. Kimport, and H. Swisher, *Quantum modular forms and singular combinatorial series with repeated roots of unity,* submitted. arXiv:1902.10698 [math.NT].

12. A. Folsom and S. Kimport *Mock modular forms and singular combinatorial series,* Acta Arith. 159 (2013), 257–297.

13. A. Folsom, K. Ono, and R.C. Rhoades, *Mock theta functions and quantum modular forms,* Forum Math. Pi 1 (2013), e2, 27 pp.

14. M. I. Knopp, *Modular functions in analytic number theory,* Markham Publishing Co., Chicago, Ill., 1970.

15. R. Lawrence and D. Zagier, *Modular forms and quantum invariants of 3-manifolds,* Asian J. Math. 3 (1) (1999) 93–107.

16. K. Ono, *Unearthing the visions of a master: harmonic Maass forms and number theory,* Current developments in mathematics, (2008), 347-454, Int. Press, Somerville, MA, (2009).

17. H. Rademacher, *Topics in analytic number theory,* Die Grundlehren der math. Wiss., Band 169, Springer-Verlag, Berlin, (1973).

18. D. Zagier, *Ramanujan's mock theta functions and their applications (after Zwegers and Ono-Bringmann),* Séminaire Bourbaki Vol. 2007/2008, Astérisque 326 (2009), Exp. No. 986, vii-viii, 143-164 (2010).

19. D. Zagier, *Quantum modular forms,* Quanta of maths, 659-675, Clay Math. Proc., 11, Amer. Math. Soc., Providence, RI, 2010.

20. S. Zwegers, *Mock theta functions,* Ph.D. Thesis, Universiteit Utrecht, 2002.

21. S. Zwegers, *Multivariable Appell functions and nonholomorphic Jacobi forms,* Res. Math. Sci. 6 (2019), 16.

Printed in the United States
By Bookmasters